Biorefineries - Selected Processes

Edited by Krzysztof Biernat

Published in London, United Kingdom

IntechOpen

Supporting open minds since 2005

Biorefineries - Selected Processes
http://dx.doi.org/10.5772/intechopen.94654
Edited by Krzysztof Biernat

Contributors
Klara Birikh, Matti Heikkilä, Alex Michine, Petri Ihalainen, Penghua Qiu, Xiye Chen, Linyao Zhang, Li Liu, Chang Xing, Yan Zhao, Roger Ruan, Kirk Cobb, Deepak Sharma, Edson Talamini, Antonio Luiz Fantinel, Rogério Margis, Homero Dewes, Amit Kumar Bajhaiya, Arathi Sreenikethanam, Krzysztof Biernat, Paulina Luiza Grzelak

First published in London, United Kingdom, 2022 by IntechOpen
IntechOpen is the global imprint of INTECHOPEN LIMITED, registered in England and Wales, registration number: 11086078, 5 Princes Gate Court, London, SW7 2QJ, United Kingdom
Printed in Croatia

British Library Cataloguing-in-Publication Data
A catalogue record for this book is available from the British Library

Additional hard and PDF copies can be obtained from orders@intechopen.com

Biorefineries - Selected Processes
Edited by Krzysztof Biernat
p. cm.
Print ISBN 978-1-83969-734-0
Online ISBN 978-1-83969-735-7
eBook (PDF) ISBN 978-1-83969-736-4

For EU product safety concerns:
IN TECH d.o.o., Prolaz Marije Krucifikse Kozulić 3, 51000 Rijeka, Croatia, info@intechopen.com or visit our website at intechopen.com.

Meet the editor

Krzysztof Biernat, Ph.D. (MechEng), is a professor at the Łukasiewicz Research Network - Automotive Industry Institute, Poland. He is also the chairman of the Polish Biomethane Council, a member of the Management Board of the "Bioeconomy Cluster" Association, and a member of the Steering Committee of the European Technology Platform Renewable Heating and Cooling. He specializes in chemical thermodynamics, including environmental processes and production technology, and quality assessment and use of operating fluids such as biofuels and alternative energy carriers. He has many national and international awards, decorations, and orders for scientific and pro-innovative activities. He is a leading expert in the International Renewable Energy Agency (IRENA), and an expert in national and European operational programs. He is the author of more than 250 peer-reviewed publications and a dozen or so monographs on the properties and operating conditions of fuels, biofuels, and other operating fluids, as well as environmental protection. He has promoted more than 200 master's and engineering theses. He is also the chairman of the Scientific Council of the journal *Material Engineering* and the thematic editor of *Applied Sciences*, *Chemical Industry*, and *Automotive Archives*. He is a member of the Academy of Engineering in Poland and a member of many national and foreign scientific societies, including the American Chemical Society (ACS), the Society of Chemical Industry (SOCI), and the American Association for the Advancement of Science (AAAS).

Contents

Preface

Biorefinery systems are one of the basic ways to mitigate the negative effects of the functioning of local ecosystems by converting biomass and organic waste into various substances such as chemicals or biomaterials and energy carriers. The research carried out so far has shown the effectiveness of waste biomass processing by obtaining the so-called added value in products and minimizing the amount of naturally generated or produced waste substances. Biorefinery systems are industrial open systems where biomass, other waste substances, and energy streams flow into these systems as inputs. Inside the system, there are several processes that result in, among others, energy exchange of the system with the environment in the forms of heat and work. The output streams of biorefinery systems include several products, such as fuels, chemicals, high-value chemicals obtained in small amounts, low-value chemicals obtained in large amounts, feed and food products, polymers, and other materials, as well as processed energy in cogeneration or trigeneration (heat, electricity, and cooling) and process wastes. An additional advantage of the functioning of biorefinery systems is the possibility of further use of the generated waste substances in subsequent technological processes. Hence, biorefinery systems should be a tool for the implementation of sustainable development in the processes of energy use of natural and waste resources. According to the concept of sustainable development, this type of energy-generating installation is the most optimal solution, which at the same time considers continuous technological development and the production of so-called clean energy and other products while reducing emissions of greenhouse gases and harmful compounds. It is a low-waste technology that uses the existing potential of waste biomass, which is currently not used at all or is used in an unreasonable way.

The basic biorefinery processes used after the pretreatment of biomass material include enzymatic hydrolysis and fermentation, including enzymatic hydrolysis, rapid pyrolysis, and hydrothermal processing (HydroThermal Upgrading [HTU]), called liquefaction or hydrothermal pyrolysis, with a possible further hydrodeoxidation (HDO) process. Currently, there are two basic technological concepts for the implementation of biorefinery processes. The first includes systems in which value-added products such as biochemicals, including biopolymers, are recovered, and the post-process residues are processed into energy carriers for internal use. The second involves the processing of waste raw material into various types of energy carriers, and in the second stage, further processing of the residues into products with added value.

Considering the progress in research aimed at developing technologies that can be used in biorefinery processes, the intention behind publishing this book is to examine the current state of research and the prospects for implementing the results of these studies in the construction of biorefinery systems fully complying with the principles of sustainable development.

Chapter 1 discusses selected environmental conditions for the functioning of biorefineries and biorefinery systems and their impact on the circular economy.

It characterizes the basic indicators for assessing the level of sustainable development of biorefinery systems, which enable these systems to fit into the circular economy. The chapter also mentions the concept of petrosynthesis as a process enabling the use of industrial waste substances in the production of conventional energy carriers.

Chapter 2 describes the state of research on the possibilities of using simulated technologies in biorefinery systems and other industrial processes like the simulated moving bed (SMB), the use of which enables effective purification of commercial end products with a higher level of purity and greater process efficiency. The chapter presents several examples of the use of the discussed technology.

Chapter 3 presents an overview of the possibilities of using enzymes in major biorefinery processes, enabling effective and efficient bioconversion processes. It discusses the enzymatic resources existing in the environment and analyzes the processes of biomass hydrolysis, lignin valorization, and glucose conversion.

Chapter 4 examines progress in the co-pyrolysis of coal and biomass. It summarizes the work on the mechanisms of synergy of co-pyrolysis processes, including the migration of alkali metals and alkaline-earth metals (AAEM). It also characterizes the influence of these metals on the basic properties of co-pyrolysis products. Finally, the chapter discusses prospects for further development of the discussed technology.

Chapter 5 summarizes the available research results on the processes of obtaining bioplastics from algae raw material. It characterizes the required properties of bioplastics and demonstrates the potential of algae and cyanobacteria as raw materials for their production. The chapter also discusses basic achievable biopolymers and the prospects for the production of biopolymers in the future, also based on biorefinery technologies.

Finally, Chapter 6 assesses the technological advances in synthetic biology and their use in the production of ethanol from cellulose raw material. The chapter presents a broad analysis of patented solutions in this area based on extensive bibliographic sources.

The chapters presented in this monograph do not cover all issues related to the state of research in the field of further development and industrial implementation of biorefinery systems, but their selection is characterized by the level of development of current research and analyzes leading to the development of comprehensive industrial technologies characterized by a high degree of sustainability, at the same time leading to obtaining both products as well as energy carriers.

I would like to thank Ms. Paulina Luiza Grzelak, MSc, for her invaluable substantive and editorial help in developing this monograph. I hope that Ms. Paulina's participation in the editing of this monograph will contribute to her further scientific development.

On behalf of myself and the authors, I would like to thank Ms. Karmen Đaleta from IntechOpen for her patience, understanding, and, most importantly, effective coordination of the complex publishing process of this book.

I dedicate this book to my former son-in-law Grzegorz Chojnacki, a recognized producer of brand-name food alcohols, with the hope that he will expand his production potential with the processes of using post-production substances in a biorefinery system implemented for his own needs.

Krzysztof Biernat
Professor-Area Lider,
Łukasiewicz R&D Network – Automotive Industry Institute,
Warsaw, Poland

Chapter 1

Introductory Chapter: Environmental Conditions of the Functioning of Biorefineries and Biorefinery Systems and Their Impact on the Circular Economy

Krzysztof Biernat and Paulina Luiza Grzelak

1. Introduction

According to the assumptions contained in many scientific studies and reports, biorefinery systems should constitute a permanent element of the circular economy, because an important feature of these systems is the full use of various types of substances, mainly waste organic substances and their environmentally friendly processing into energy carriers and other products with added value. The basic diagram characterizing the basic value chain in the bioeconomy, consisting of the use of biorefinery systems, was developed already in 2013 and presented in **Figure 1**.

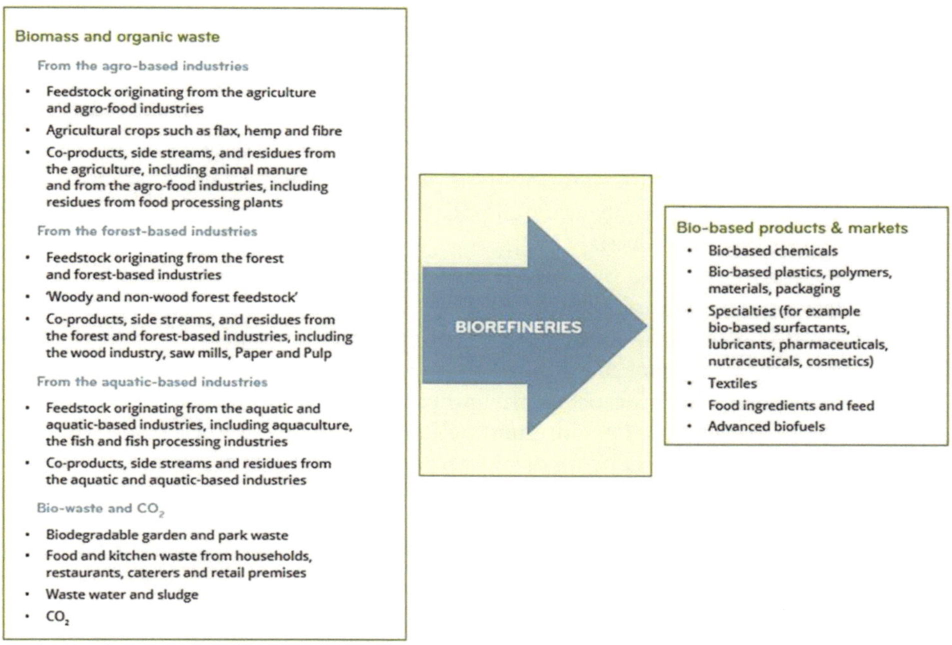

Figure 1.
Value chains in the bioeconomy [1].

The possible processes of processing waste raw materials through biorefinery systems shown in **Figure 1** should be considered desirable and consistent with the principle of sustainable development. This compliance can be determined by monitoring these changes with the use of tools such as measures or indicators of sustainable development. These tools allow to determine the pace and scope of changes, as well as the effectiveness of the activities carried out. The measures are also intended to verify and objectify the goals set by the international community and individual countries [2].

2. Assessment indicators of the sustainable development in biorefinery systems

One of the first international organizations which attempt to develop a system of measures was the United Nations European Commission of Economy—UNECE. It developed a system of twelve measures, which were also assigned to twelve environmental problems. However, they did not fully reflect the complex and interdisciplinary nature of sustainable development.

Only the list of measures presented in the report "An Overview of Environmental Indicators, State of the Art and Perspectives", developed under the United Nations Environment Program (UNEP), can be considered comprehensive and allows for the assessment of various aspects of sustainable development [3], as well as the measures included in two reports elaborated by the United Nations Commission on Sustainable on Sustainable—UNCSD:

- Indicators of Sustainable Development: Methodology Sheets (April–May 1996);
- Indicators of Sustainable Development: Framework and Methodologies (August 1996).

As part of the UNCSD program for the sustainable development indicator systems, a list of 130 indicators compiled under the cause-state-reaction scheme was developed [4].

On the cause side, one should look for those forms of human activity that directly or indirectly affect the sustainability of development. State indicators define the implementation of sustainable development, while response indicators show human activities (e.g. changes in environmental policy and others) in response to the challenges of sustainable development) [2].

The indicators developed by UNCSD are to be used in the environmental policies of individual UN member countries. They also provide a kind of background for the definition of national indicators. In order to facilitate this task, UNCSD has prepared a list of methodological recommendations for their construction, including, inter alia, the reference of the indicator to the Agenda 21 objectives or the relevance to the sustainable development policy [4].

The hope placed on the concept of biorefinery installations as an element contributing to the creation and ensuring of sustainable development made it necessary to introduce sustainability measures for the individual planning and design stages of such installations. Their complexity and multifaceted nature led to the derivation of a number of economic, environmental, and social indicators in this regard. The proposed indicators are often characterized by similarities as well as divergences, but they become useful when they are used as complementary measures. The system of indicators is constantly evolving, growing, and new methodologies are being proposed.

The environmental impact of biorefineries is significant in terms of climate change, eutrophication, water scarcity, land use, energy, and material depletion, as well as water pollution and toxicity. Several aspects can be categorized according to the type of biorefineries. Therefore, among a number of indicators characterizing sustainable development, one can list those that will be used in the analysis of the impact of biorefinery installations on the environment. These are the indicators:

- Process efficiency (energy conversion efficiency),
- Use of primary resources (accumulated primary energy, abiotic resources, land, clean water),
- Environmental impact (GHG emissions to the atmosphere, water eutrophication, soil acidification, potential for the formation of photochemical oxidants, emissions of toxic substances to water reservoirs and watercourses),
- Economic efficiency (income potential of biorefineries).

In the indicator methodology proposed by Sacramento-Rivero [5] sustainable development is mainly understood as a state of efficiency of a biorefinery in which the consumption of resources and the associated effects are at such a level that the biorefinery can operate successfully for an indefinite period with a limited environmental impact. This means that current human needs (those corresponding to the activities of biorefineries) are met without compromising the ability of future generations to meet their needs. It also means that the good performance of the assessment today may not be as good in the future if conditions change, for example if the legislation tightens the permitted emission levels or if agricultural yields start to decline.

A biorefinery is assumed to have a number of points where it is, is not sustainable to some degree. On this scale "0" represents the ideal state of sustainability, and "1" is the critical limit beyond which the installation becomes unsustainable. Consequently, the scale takes real values from zero to infinity, while sustainable development takes place with values from zero to one. The closer to zero, the more sustainable development is, while the closer to "1", the more sustainable development decreases. Values greater than one mean unsustainable development (**Figure 2**).

In this normalized sustainability scale, values from zero to one represent balanced states, while values equal to or greater than one represents unsustainable system operation. The state of ideal durability, in the conventional "0" point, is purely conceptual - it represents the ideal value of the indicator, in the absence of environmental impact (zero-emission), impossible to achieve in reality.

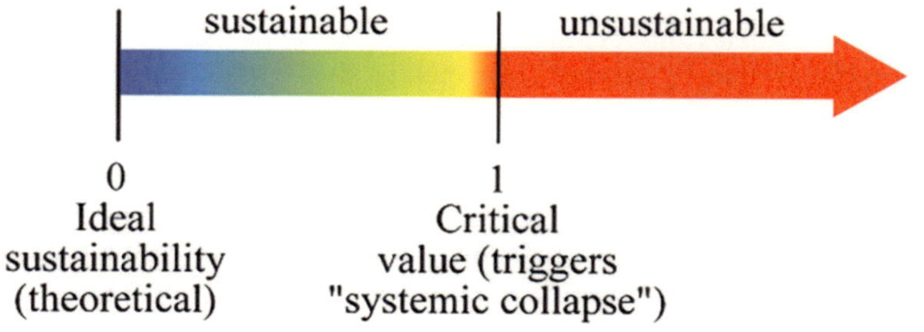

Figure 2.
Sustainability scale—range of states of sustainability [5].

On the other hand, the critical value is the point "1" at which the system becomes unbalanced in the sense that it represents an excessive negative environmental impact relative to the reference system (e.g. conventional oil refinery) or is uneconomical. The "1" limit point is a moving point that depends on the ideal durability indicator and the critical value based on various factors, incl. Reference to national and international targets (e.g. reduction targets). For some indicators, this critical value will require periodic recalibration.

The critical limit value must be reliably established determined on a solid scientific basis and confirmed by national or international targets, agreements, or certification bodies. The selection of the appropriate critical value will depend on the results of the Life Cycle Assessment (LCA) of the installation and its products and the sustainability assessment taking into account the definition of the indicator. When at least two sources propose different critical values, the general recommendation is to choose the most stringent criterion, in line with the precautionary principle.

Using the example of the indicator: "Reduction of greenhouse gas (GHG) emissions", an ideal system would not emit greenhouse gases (this state would be described by a zero value on a normalized scale), and the system at the critical limit point would be one that releases exactly the amount of greenhouse gases, which the atmosphere is able to take without reaching a catastrophic rise in temperature. Due to the complexity of determining the above, the critical reduction of greenhouse gas emissions would be the one required by the appropriately adopted target (e.g. the reduction target), the benchmark should be the results achieved by the reference system. Therefore, if a biorefinery is designed for a country belonging to the European Union (EU), then the reduction targets, and thus the critical (border) reduction value, will be determined on the basis of the currently applicable RED II (Renewable Energy Directive) on renewable energy sources. In the RED II Directive [6], the level of reduction of greenhouse gas emissions is min. 65% for installations launched after 01/01/2021 and min. 70% for renewable fuels of nonbiological origin. In this case, where two objectives apply to the designed biorefinery, the critical value should be established according to more stringent regulations.

3. Sustainable development criteria in the field of energy carriers as biorefinery products

The European RED Directive establishes sustainability criteria, which were then revised and tightened up in RED II in Article 29 (1). 2–8 and 10. They read as follows:

- Biofuels produced must not come from raw materials obtained from land with high biodiversity (i.e. primeval forests and other wooded lands, protected areas, grassland with high biodiversity),
- Produced biofuels cannot come from raw materials obtained from land rich in carbon (i.e. wetlands, wooded areas, peatlands),
- Agricultural raw materials grown in the EU and used for the production of biofuels should be obtained in compliance with minimum requirements concerning good agriculture culture in accordance with environmental protection and with certain statutory management requirements set out in the Common Agricultural Policy,
- The greenhouse gas emission savings through the use of biofuels, bioliquids, and biomass fuels should be at least 50%, 60%, and 65% respectively for

biofuels, biogas consumed in the transport sector, and bioliquids produced in installations put into operation before October 5, 2015, before December 31, 2020, and after January 1, 2021.

Although biofuels are very important from the point of view of reducing greenhouse gas emissions in EU countries, it should be borne in mind that the production of currently known and used biofuels usually takes place using arable land that was previously used for other agricultural purposes, such as food or feed production. These are the so-called first-generation biofuels (conventional biofuels). Since agricultural production for food is and will remain necessary, it can lead to the expansion of food crops to land that has not been cultivated so far, thus including high-carbon areas such as forests, wetlands, and peatlands. Indirect Land Use Change (ILUC)—as the above-described process is called—can cause the release of CO_2 stored in trees and soil, which may counteract the process of reducing greenhouse gas emissions resulting from the increase in the share of biofuels. To tackle the ILUC issue in the "Clean Energy for All Europeans" package [7], the revised Renewable Energy Directive introduces a new approach. It sets limits for biofuels, bioliquids, and biomass fuels with a high ILUC risk, with significant expansion in areas with high carbon content. These limits will have an impact on the amount of these fuels that the Member States can count toward meeting their national targets when calculating the total national share of renewables and the share of renewables in transport. EU Member States will still be able to use (and import) fuels that fall under these limits, but will not be able to take these amounts into account when calculating the extent to which they have met their renewable energy targets. The limits in question are to be frozen in the period 2021–2023 at the 2019 levels, and then until the end of 2030, the limit is gradually reduced to 0%.

RED II also introduces an exemption from these limits for biofuels, bioliquids, and biomass fuels certified as having a low ILUC risk. In order to implement this approach, as required by the Directive, the Commission has adopted Delegated Regulation (EU) 2019/807.

According to the current target by 2030, at least 14% of transport fuels are to come from renewable sources, with first-generation ILUC-risk biofuels no longer counting toward the EU's renewable energy targets from 2030. As a part of this goal has been set a target for advanced biofuels produced from feedstocks listed in Part A of Annex IX. The share of advanced biofuels and biogas produced from raw materials listed in Part A of Annex IX to the Directive (biofuels and biogas with low ILUC risk) as a share of final energy consumption in the transport sector will be at least 0.2% in 2022, at least 1% in 2025 and at least 3.5% in 2030.

Member States of the European Union may exempt fuels suppliers that supplying fuels in the form of electricity or renewable liquid and gaseous transport fuels of nonbiological origin—with regard to these fuels—from the requirement to achieve a minimum share of advanced biofuels and biogas produced from raw materials listed in Annex IX, Part A.

Raw materials for the production of biogas for transport and advanced biofuels, the share of which in the minimum shares referred to in Art. 25 Section 1, the first and fourth paragraphs, can be considered as twice their energy value.

The share of biofuels and bioliquids, as well as biomass fuels used in transport, if produced from food and feed crops, may not be higher than by one percentage point than the share of these fuels in final energy consumption in transport and rail transport sectors in 2020, with a maximum of 7% of energy consumption in the road and rail transport sectors in Member State.

Fuels produced from high-risk ILUC raw materials will be constrained by a stricter consumption cap in 2019. The share of biofuels, bioliquids, or biomass fuels with a high indirect land-use change risk, produced from food and forage crops, where the significant expansion of the production area into carbon-rich land shall not exceed the consumption of such fuels in the Member State in 2019, unless they are certified as biofuels, bioliquids or biomass fuels with a low risk of indirect land-use change. From 31 December 2023 to 31 December 2030 at the latest, this limit will be gradually reduced to 0%.

In relation to the changes introduced by the currently applicable RED II directive, it is reasonable to use raw materials with a waste status in biorefinery production from various industries (agriculture, food, livestock, etc.), catering waste, or, for example, out-of-date food products. The use of this type of raw material is also supported by a significant reduction in emissions in the life cycle of the biorefinery system, resulting from the avoidance of the need to use the land for the cultivation of the raw material and all processes related to sowing, harvesting and transporting the raw material to the biorefineries.

Therefore, biorefinery installations will undoubtedly contribute to the more ambitious target of a 55% reduction in greenhouse gas emissions by 2030 [7], forecasted in the Climate Goals Plan [8] in the frame of the European Green Deal [9].

4. Conclusions

In the light of the latest regulations, the indicators analysis for biorefineries comes down to the calculation of the emission factor for the full life cycle of a biorefinery product, which is a biofuel/biocarbon biocomponent. For chemicals and other value-added products, comparisons can be made by assessing the LCA of non-biorefinery product equivalents. The emission factor defines the possibility of using a fuel/biofuel or a product with an added value on the market, and thanks to this indicator it can be assessed whether this fuel or product actively participates in the achievement of the reduction target. However, it should be noted that in 2021

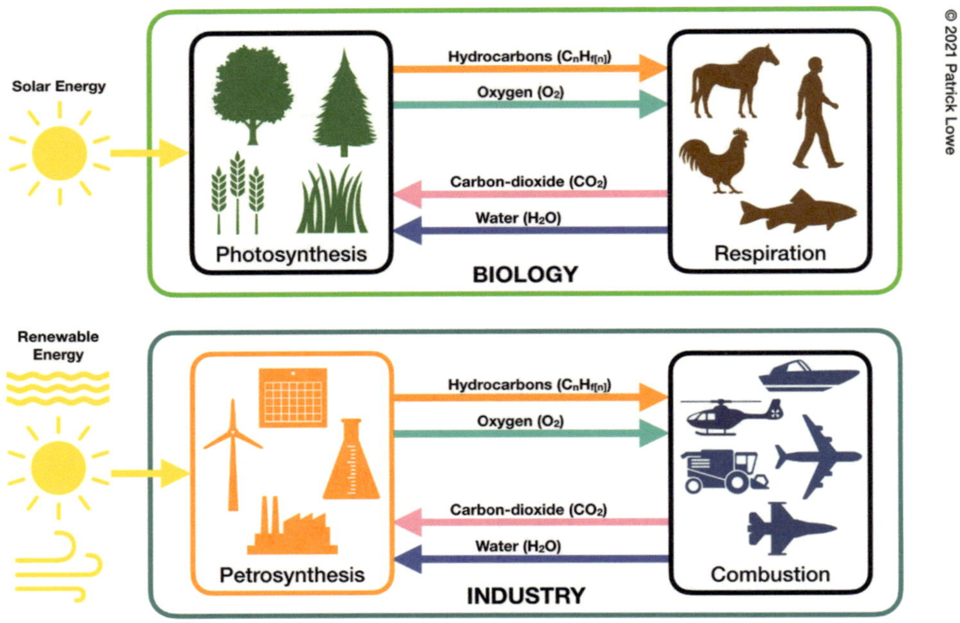

Figure 3.
Simplified models of petrosynthesis and photosynthesis [10].

the concept of petrosynthesis [10] appeared as a series of complex technological processes using mainly the final products of combustion, such as carbon dioxide and water (water vapor), and other greenhouse gases using energy from renewable sources generating electricity, to the production of hydrocarbons and their compositions which are equivalent to products manufactured so far by the refining and petrochemical industries. According to its author, this concept is a practically possible substitute for photosynthesis processes, as it was shown in **Figure 3**.

The concept presented in **Figure 3** may partially complement the processes carried out by comprehensive biorefinery systems, allowing for the full use of all possible post-process waste, including water and carbon dioxide, in the so-called closed cycle. However, the actual implementation of this concept requires further work, especially in the field of obtaining waste carbon dioxide from dispersed processes and hydrogen directly from industrial, municipal, and salt waters, as well as from other sources, which is related to the increasing shortages of water necessary for biological processes, in including drinking water.

Author details

Krzysztof Biernat[1]* and Paulina Luiza Grzelak[2]

1 Łukasiewicz R&D Network – Automotive Industry Institute, Warsaw, Poland

2 Motor Transport Institute, Warsaw, Poland

*Address all correspondence to: krzysztof.biernat@pimot.lukasiewicz.gov.pl

References

[1] Biobased and Renewable Industries for Developmnent and Growth in Europe. Brussels: Strategic Innovation and Research Agenda; 2013

[2] Borys T. Wskaźniki Ekorozwoju (Indicators of Ecodevelopment). Białystok: Wyd Ekonomia I Środowisko; 1999

[3] An Overview of Environemntal Indicator, State of the Art and Perspectives. New York: RIVM Publications Repository; 1994

[4] Indicators of Sustainable Development: Metodology Sheets, Background Paper No. 15. New York: Department for Policy Coordination and Sustainable Development, United Nations; 1996

[5] Sacramento-Rivero JC. A methodology for evaluating the sustainability of biorefineries: 5 framework and indicators, biofuels. Bioproducts and Biorefining. 2012;**6**:32-44

[6] RED II Directive. Directive (EU) 2018/2001 of the European Parliament and of the Council of 11 December 2018 on the promotion of the use of energy from renewable sources (recast) (Text with EEA relevance)

[7] European Commision. Clean Energy for all Europeans. Luxembourg: Publications Office of the European Union; 2019

[8] Communication from the Commission to the European Parliament, the Council, the European Economic and Social Committee and the Committee of the Regions. A more ambitious 2030 climate target for Europe. In: Investing in a Climate Neutral Future for the Benefit of Citizens. Luxembourg: COM; 2020 562 final

[9] Communication from the Commission to the European Parliament, the Council, the European Economic and Social Committee and the Committee of the Regions. The European Green Deal. Luxembourg: COM; 2019 640 final

[10] Lowe P. White Paper – Petrosynthesis – The Competition of the Industrial Revolution. © 2021 Patrick Lowe, 15 November 2021. Available from: https://zeropetroleum.com/wp-content/uploads/2021/11/White-Paper-Petrosynthesis-15-11-21.pdf

Chapter 2

Simulated Moving Bed Technology: Overview and Use in Biorefineries

Deepak Sharma

Abstract

Synthesis of chemical compounds oftentimes produce a mixture of desired and undesired components. The ease of purification and recovery of the desired component more often than not determines the viability of the production technology. Simulated Moving Bed (SMB) technology is a continuous purification and separation technique with better performance (less solvent consumption and higher throughput) than traditional batch chromatography. SMB is a continuous separation technology which can be used to achieve the desired product purity with considerably lower power and raw material consumption. A lot of research and development is undergoing in the SMB technology which is enabling the search for more economical and carbon neutral ways of producing industrial chemicals. SMB has proven to be of great assistance in extracting products produced in biorefinery fermentation processes in an economical and energy efficient fashion. This chapter outlines the various processes the author has developed using SMB technology, its use in biorefineries, and prospective use in the future.

Keywords: Simulated Moving Bed, Chromatography, Adsorption, SMB, Purification, Separation, Unit operation, Manufacturing, Adsorbent, Desorbent, Extract, Raffinate, Biorefienry

1. Introduction

Simulated Moving Bed (SMB) technology is an improvement over the traditional batch chromatography to a continuous chromatography process. SMB allows scale-up of lab chromatography process for high production outputs needed in the chemical industry. Most chemical reactions also produce many byproducts along with the desired product. Therefore, many purification steps are needed in order to achieve the desired purity of the final product. Synthesis of products have traditionally been given more importance in the chemical industry. The recovery and purification of desired product is often more time consuming and costly in comparison to the synthesis reaction. There are several widely employed purification techniques, such as distillation, crystallization, filtration etc. Almost every chemical manufacturing operation requires the use of purification process to achieve the desired product purity. Oftentimes the choice and efficiency of the selected purification technology determines the quality and cost of the product. Liquid-phase adsorption has traditionally been used in the industry to remove specific by-products produced during synthesis of industrial chemicals. Batch chromatography used for purification of feed streams use principles such as adsorption, size exclusion, complexation, ion-exchange, hydrogen bonds or a combination of these mechanism. SMB chromatographic technique was first used in the petrochemical industry in the 1940s [1]. In this chapter the author focuses on use of SMB in the energy, pharmaceutical, and nutraceutical industry and how the use of this technology enables the development of greener and sustainable process.

2. Background

Batch chromatography has been used in the industry for several years, however, the technology has suffered from low yield, high solvent consumption, poor column utilization and high product dilution. SMB technology was developed to overcome the limitations of batch chromatography and convert it into continuous process. The very nascent form of SMB technology dates all the way back to the 1840s where fixed beds and moving ports to simulate a counter current movement of liquids and solids was first used in Shank's system for leaching [2]. A family of SMB processes were developed by UOP LLC (Des Plaines, Illinois, U.S.) for the petrochemical industry. Some of the processes industrialized by UOP were Parex for the recovery of para-xylene, Molex for the separation of linear paraffins, Ebex for the separation of ethyl benzene from a mixture of C8 aromatic isomers, Olex for the separation of paraffins and olefins. Until 1970 most of the processes for SMB technology were in the petrochemical industry. The sugar industry started seeing SMB process development around 1980s when UOP developed the Sarex sugar purification process [3, 4]. In 1990s the use of SMB technology in pharmaceutical industry caught speed. Several processes were developed using chiral stationary phase [5–7]. SMB technology has found its use in almost all chemical industries including pharmaceutical, petrochemical, waste removal, enzyme separation, organic acid separation, purification of fine chemicals. Traditionally SMB technology has only been used for binary separations i.e. two component or at most three components. Most recently there has been developments in SMB technology to separate multiple components from a complex feed stream. The author has developed several complex SMB processes that are outlined in this chapter.

3. Separation principles for batch chromatography and SMB

In batch chromatography a column is packed with adsorbent material which is called the stationary phase or sorbent. The stationary phase is determined based on the characteristics of the feed stream. Typical stationary phase used are carbon, silica gel, alumina, modified silica etc. Once the column is packed it is flushed with a mobile phase. Mobile phase is a single solvent or a mixture of solvents that are decided based on the solute solvent interactions. Similar to analytical chromatography the feed solution is injected into the column and then eluted with the mobile phase called desorbent. Mobile phase and desorbent will be interchangeably used in this chapter. The components of the feed that have higher affinity for the sorbent are referred to have a higher partition coefficient. Having a higher partition coefficient means that the component is strongly bound to the sorbent so has a higher concentration on the sorbent than in the mobile phase. Since only the components in the mobile phase migrate with the desorbent a higher affinity feed component migrates slower than a low affinity feed component thereby creating a concentration gradient in the column resulting in separation of the various feed components.

Batch chromatography can be used to separate such components by collecting the desired product in one vessel and reject byproducts to a separate vessel. In most industrial applications there are components in the feed that either have a higher affinity or lower affinity than the desired product. A higher degree of separation is desired from the two undesired components to achieve a pure product. In an ideal world one would assume that one can easily achieve this in a single column with a reasonable column length. In practice this is not feasible as the short column length does not give enough run length to the components, as a result of which both purity and yield is compromised. To overcome this drawback of single column separation, SMB technology was developed to provide the much-desired solution.

SMB technology overcomes most of the shortcomings of batch chromatography. In a SMB several identical columns are connected in series. **Figure 1** shows eight columns connected in series such that the outlet from one column is connected to the inlet of the next column. In this figure the desorbent enters the first column and feed enters the fifth column. The feed in this illustration contains two components A and B. The fast-moving component A is removed in the raffinate stream which comes out from the bottom of column # 7. The component B, which is strongly adsorbed on to the stationary phase, is tapped from the bottom of column # 3, labeled as extract. The inlet and outlet ports divide the eight-column setup connected in series into four zones. Each zone in a SMB has different flow rate. The ports move periodically along the loop to follow the migrating solute bands. On every switch of SMB port, the desorbent and feed move one column downstream to the inlets of column 2 and column 7 respectively. Even the extract and raffinate port move one column downstream and are drawn from column 3 and column 8 respectively.

The inlet and outlet streams determine the various zones in the SMB. The zone between the desorbent inlet and extract outlet is zone I which is called the desorption zone. The zone between the extract and feed port is zone II which is called the

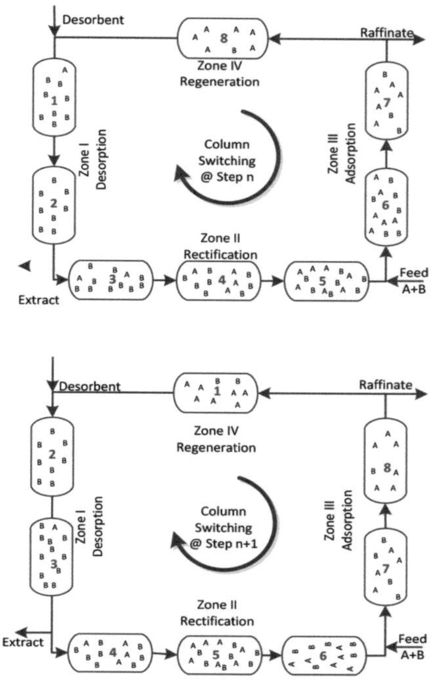

Figure 1.
SMB setup with column switch illustration.

rectification zone. The zone between the feed port and the raffinate port is zone III which is called the adsorption zone. The zone between the raffinate and desorbent is zone IV which is called the regeneration zone. The column switch time is determined carefully such that the slow-moving feed component B never reaches the raffinate port and the fast-moving component A never reaches the extract port. Fine tuning the switching time ensures the fast-moving component A only come out from the raffinate port and the slow-moving feed component B only comes out of the extract port. The setup can be optimized to achieve the desired product purity and yield. This periodic port switching achieves a simulated counter current movement of the solid phase with respect to the mobile phase. This enables a continuous process where the desired product can be continuously produced.

4. Advantages of SMB technology

There are several advantages of using SMB over batch chromatography or any other industrial purification technologies. One of the advantages of SMB is the continuous purification of feed components achieved at high product purity and yield. SMB also considerably reduces the amount of desorbent usage over batch chromatography. The continuous regeneration of the adsorbent in a single setup also reduces the usage of adsorbents considerably thereby increasing the overall adsorbent utilization. The reduced consumption of desorbent [8] and adsorbent makes this a greener process as it enables in reducing the carbon footprint of achieving the same purification with higher efficiently. These improvements also lend its advantage to cost savings not only in operating cost but also in capital cost by requiring a smaller footprint, equipment size, and manpower. More importantly, SMB technology is particularly useful in biorefineries as it is naturally favorable for the low concentration product stream from biorefineries. Fermentation reactors are typically limited by the concentration of product one can achieve in the fermenter before the producing organisms get deactivated. The low concentration is favorable for adsorption, making SBM the preferred technology to achieve economic purification of the products.

Biorefineries particularly face a huge challenge in isolating the desired products from the reaction mixer. The cost of separation and purification account for a large portion of costs in biorefineries. SMB provides the much-required cost effective solution which makes biorefinery production feasible and competitive to tradition synthetic production techniques.

5. Shortcomings of SMB technology

In the past decade a lot of technological development has happened in the field of SMB technology. The author has enabled the commercialization of some of the purification processes. An SMB process involves a complex design which requires a lot of inputs. A typical 4 zone SMB requires research on multiple lab-scale experiments to determine the flow rate of feed, desorbent, extract, and raffinate. Along with that the frequency of switching the columns need to be determined, the zone length for each of the four zones, and the operating temperature. A total of ten design parameters need to be identified for the successful run of SMB setup.

During startup of SMB processes all the ten parameters need to be developed using experimental setups and scaled up to parameters for the manufacturing unit. Every run of the SMB setup involves optimization which leads to production of off spec product. Even slight variation in any parameter can easily disturb the equilibrium of the entire process. Moreover, some parameters drift and are heavily dependent on accuracy of pump flow rate and varying pressure drop in the different columns.

6. SMB operation

In this setup the author will compare SMB process with a hypothetical moving-bed system shown in **Figure 2**. In the schematic the desorbent D moves from the top to bottom counter current to the adsorbent which moves from bottom to top. The desorbent chosen is a solute which has a lower boiling point than the feed components. This setup explains separation of a binary mixture A and B. Component A is the faster moving component and does not interact with the adsorbent as much as B does. Component B is adsorbed selectively on to the adsorbent. As shown in the **Figure 2** feed enters the column at a particular point on the column. The desorbent moves the faster moving component A which has little interaction with the adsorbent with it and comes out with the desorbent from the raffinate which is right below the feed point. The adsorbent moves the component B due to increased interaction with the adsorbent and come out at the extract port right above the feed.

The inlet and outlet of streams split the entire column into 4 zones:

Zone 1: This is the desorption zone. The zone between extract and desorbent is called the desorption zone. In this zone component B is removed from the adsorbent (solid).

Zone 2: This is this rectification zone. The zone between the feed and extract is called the rectification zone. In this zone component B is selectively enriched on the adsorbent over component A.

Zone 3: This is the adsorption zone. The zone between the raffinate and feed is called the adsorption zone. In this zone component B is adsorbed on the adsorbent (solid).

Zone 4: This is the regeneration zone. The zone between the desorbent and raffinate is called the regeneration zone. This zone acts as a buffer to prevent component B from reaching the raffinate port.

In practice there are two types of continuous chromatography operations prevalent in the industry. Moving bed operation [9] where the actual adsorbent bed

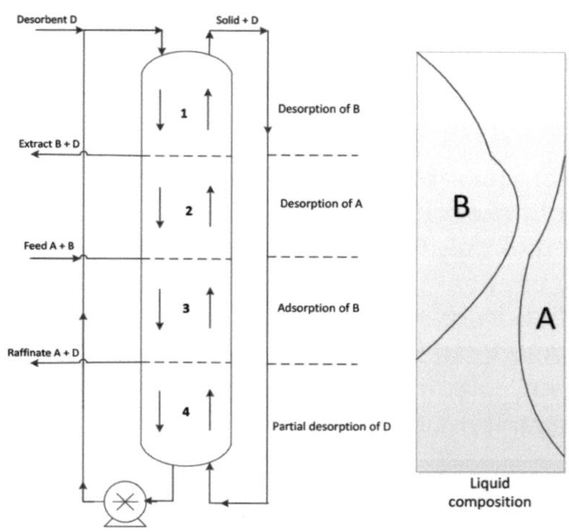

Figure 2.
Illustrating SMB as a distillation column.

moves and another one is a simulated moving bed [10] where the movement of the beds is simulated by series of valves [11, 12].

7. SMB parameters

7.1 Adsorbate-adsorbent interaction

Selection of separation technology employed for any particular application is determined by the phase relation that can be developed using various separative agents. Adsorption is a complex phenomenon compared to any other form of separation technique employed in the industry. Adsorption is typically used as a last resort when none of the other separation techniques are practically feasible.

Interaction of components that are intended to be purified need to have selectivity for the adsorbent that is chosen for purification. Moreover, there needs to be enough selectivity between components to achieve a higher purity and yield of the components. There is not much literature on adsorbate-adsorbent interaction. The only literature available is covered in patents [13–21]. With the little that has been established it has been concluded that the selection of adsorbent is more of an art than science.

7.2 Adsorbents used in the industry

The choice of an adsorbent for any adsorption process is primarily governed by four parameters; mass transfer rate, capacity, long-term stability, and selectivity. Adsorbents traditionally used in the industry are silica, activated alumina, and activated carbon due to their large micropore volume and surface area.

As a rule of thumb silica and alumina are used for polar components. Equilibrium data published in the literature [13–20] can be used to identify a suitable adsorbent. The order of affinity of various chemical species is; saturated hydrocarbons, aromatic hydrocarbon, halogenated hydrocarbons, ethers, esters, ketones, amines, alcohols, carboxylic acids [10]. For non-polar components the widely used adsorbent is activated carbon or modified silica. Equilibrium data is available in the literature for activated carbon [13, 14, 18–20]. Polymeric resins are mostly used in the pharmaceutical and food industry. Polymeric resins have ionic properties which have shown selectivity for cation-anion exchangers and sugar separations [21]. Resins are primarily used in aqueous phase separations.

7.3 Normal phase

Normal phase adsorbents are primarily adsorbents that are polar in nature. Polar adsorbents can either have a solid base or the solid base can be modified by chemically bonding a polar molecule to a solid base. Separations using normal phase chromatography typically use non-polar solvents or solvent mixtures with somewhat lower polarity than the adsorbent. Separations using normal phase chromatography separates molecules in order of increasing polarity. For products that are generated in biorefineries in a non-polar medium, one can use normal phase adsorbent to achieve the desired separation.

7.4 Reverse-phase

In reverse phase chromatography the adsorbent is non-polar and the solvent used is polar. Like the name suggests separation using reverse phase chromatography separates molecules in reverse order of polarity. Most commonly used adsorbents for reverse phase chromatography are silica base adsorbent bonded with trimethyl, octadecyl, cyano, butyl, octyl ligands. Reverse-phase chromatography packings find its application as both physically adsorbed-phase and bonded-phase materials, application of bonded-phase packaging have dominated the industry since the early 1980's. Fermentation products that generate products in polar medium can use reverse-phase stationary phase to achieve the separation.

7.5 Ion-exchange

As the name suggests ion-exchange separates solute molecules based on their ionic interaction with the adsorbent. Adsorbents used for ion exchange are composed of a base material like silica or a cross linked polymer primarily with a pore diameter range of 10 to 100 nm derivatized with negatively or positively charged ligands. The ligands chosen determine if the adsorbent would be a weak or strong ion-exchange resin based on the pH range they are operated in. In addition to extraction of Tagatose from a mixture of glucose and galactose from a biorefinery discussed in Section 9.1, Ion exchange has been used in biorefineries to remove inhibitors from biomass hydrolysate (acid, salts), carboxylic acid, impurities in biodiesel (glycerol, methanol, soap etc.), purification of succinic acid and many other.

7.6 Size exclusion

Size exclusion chromatography is used to separate molecules based on their molecular size. Another name for size exclusion chromatography is gel permeation chromatography. Pore size of the adsorbent determines the size of molecule the adsorbent can separate. Operating pressure of the apparatus is very critical when using size exclusion chromatography. Adsorbents used for size exclusion chromatography are carbohydrates (soft gels) or carbohydrates cross linked with agarose or acrylamides or silica 5–10 micron in diameter possessing a controlled pore size distribution.

Recently there have been several advances in the development of new sorbent materials. New adsorbents like acrylamide gels and microscopic rods of silica are being developed for chromatography application.

7.7 Desorbent selection

Selection of desorbent is equally important to the selection of adsorbent. Even though the desorbent is an inert solvent that is not a desired product, particular attention needs to be paid to the selection of the desorbent to ensure it does not compete for the adsorption sites and reduce the active sites available for the components to be separated. Another factor to consider is the boiling point of the desorbent should be somewhat different than the components to be separated so that the separated products can be easily recovered. Lower boiling point of the desorbent also favors the economics of the process. A sweet balance between safety and stripping cost should be given due consideration while selecting a desorbent. Low boiling solvents can cause a flammability hazard mitigating which can turn out to be costly for operation.

8. Lab experiments for scale-up of SMB

Every process that is commercialized and produced in the industry goes through a discovery phase. Like any other unit operation even SMB process has unique experimental setup's that can be used to determine parameters required to scale up and run industrial scale SMB setup. The author has conducted extensive research in identifying experimental setups that can be used by anyone assessing the use of SMB technology for separation of a mixture of compounds. Some of the experimental setup's that provide insight on the feasibility of the technology is outlined in this section. Every separation is unique, based on the properties of the product and medium in which the product is produced one can perform sequential experiments to identify the right adsorbent and mobile phase to achieve the separation.

8.1 Adsorbent capacity determination

The selection of adsorbent to be used for a particular separation depends on the capacity and selectivity of the adsorbent for the component in question. The capacity of the adsorbent can be determined by using the setup shown in **Figure 3**.

A column with a length of 250 mm and internal diameter 10 mm is packed with dried adsorbent for which the adsorption capacity needs to be determined. Before packing the column with the adsorbent the initial weight of the column is recorded and after filling the adsorbent with the adsorbent in consideration the final weight of the column is measured. The difference between the initial and final weight tells you the weight of adsorbent added to the column.

In order to determine the capacity of the adsorbent for a particular component, isolated pure components can be purchased from any vendor (example Sigma Aldrich) for the study. It is important that pure component is used for the study as the presence of impurity in the component can interact with the active sites of the adsorbent thereby giving a false adsorbent capacity. A 1–5% solution of the component is prepared in the mobile phase. Say if it is desired to study the adsorption capacity of ethanol on activated carbon then a 1% solution of ethanol can be prepared in water and passed through a column packed with activated carbon. To determine the adsorption of the component, the solution is passed through the column using a pump at a fixed flow rate say anywhere between 1 and 5 gpm. The effluent from the column is collected in vials. Sterilized 50 ml vials are available from several vendors for laboratory experiments. Since we do not know the capacity of the adsorbent, it is best to collect 10 ml samples of the effluent. If a rough estimate needs to be determined for the exact point of the breakthrough then the effluent from the column can be directly connected to an IR detector to determine when the peak happens. Based on the rough breakthrough volume one can decide the volume of sample required. The effluent samples collected are injected in a high-pressure liquid chromatogram to determine when the sample collected contains the

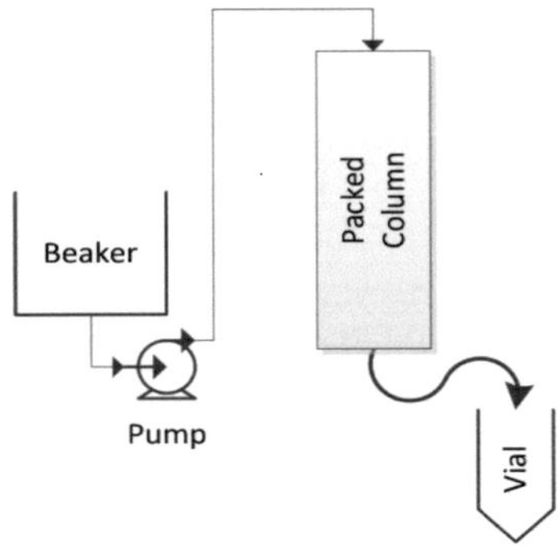

Figure 3.
Adsorption test setup.

component adsorbed on the adsorbent. It will be seen that the first few vials only contain the mobile phase in which the component is dissolved. Since the adsorbent will keep adsorbing the component till it has capacity, once the adsorbent is saturated the component will break through at which point no more adsorbent will get adsorbed on the adsorbent. The first vial which matches the composition of the feed solution marks the breakthrough point and can be called the breakthrough vial. Calculate the cumulative volume of all the vials before the breakthrough vial. Based on the cumulative volume identified one can calculate the mass of the component adsorbed on the adsorbent. Dividing the mass of component adsorbed by the mass of the adsorbent gives the adsorption capacity of the adsorbent.

$$Adsorption\ capacity\ of\ adsorbent = \frac{Mass\ of\ component\ adsorbed}{Mass\ of\ adsorbent\ in\ the\ column} \times 100 \quad (1)$$

$$Mass\ of\ componet\ adsorbed = wt\%of\ component\ in\ the\ feed\ X\ Cumulative\ volumne\ before\ breakthrough \quad (2)$$

8.2 Pulse test

Pulse test is primarily done to identify the residence time of various components as they pass through a column packed with the adsorbent used for achieving the separation in an SMB.

If one wants to conduct a quick pulse test using a lab scale SMB column (250x10mm), one can connect the discharge of the column directly to IR detector or UV detector based on the characteristics of components to be identified as shown in **Figure 4**. One can first inject a blank injection to know the void volume of the column. Blank injection can include any component that does not interact with the adsorbent and hence comes out at the void volume. For activated carbon one can use salt solution like NaCl. In the next step one can inject a solution of the component to find the residence time of the component in the IR or UV detector. Knowing the residence time of the various adsorbents can give a rough idea of starting condition of a SMB which can then be optimized.

Several applications of SMB require separation of multiple components from a feed stream. Some feed streams can contain 20–30 components that fight for the adsorbent sites. For such feed components one might need to run multiple SMB runs to get the desired component. As an example, the author has developed a process for isolating alpha-tocotrienol from palm oil [22]. Palm oil contains several

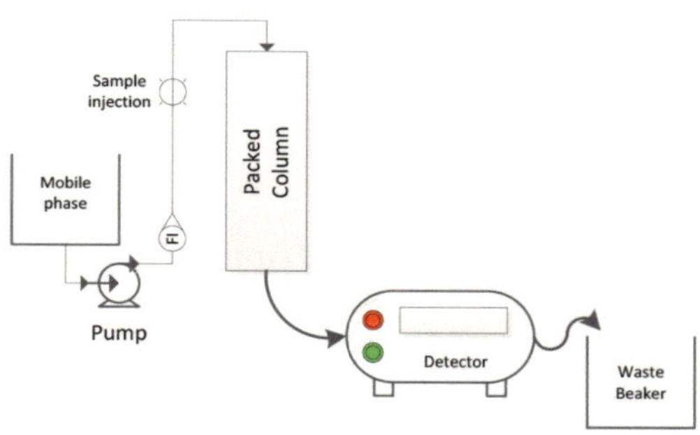

Figure 4.
Residence time detection.

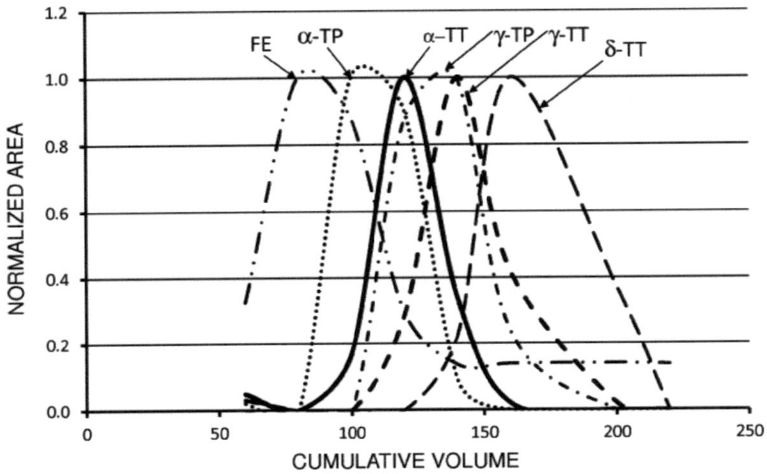

Figure 5.
Shows the results of a pulse test conducted for the palm oil extract feed. Source [22].

components like alpha-tochopherol and tocotrienol; beta-tocopherol and tocotrienol; gamma- tocopherol and tocotrienol; delta- tocopherol and tocotrienol; and several other backend and frontend components.

A detailed pulse test is required to get a good idea of residence time of various components and identify the effect of the presence of one component on other components.

In order to conduct a pulse test a stainless steel column is packed with the desired adsorbent. Tare weight of the column is registered before filling the column with the adsorbent to know the exact weight of adsorbent added to the column. Once filled the column is kept in a water bath and maintained at the operating temperature desired for the SMB run. In order to equilibrate the column the desorbent chosen for the separation is passed through the column for 30 minutes. Once the column is equilibrated a known amount of feed dissolved in the desorbent is fed to the top of the column preferably 5 ml solution of 5% feed dissolved in the mobile phase. Once the feed is injected it is then eluted with the desorbent at a desired flow rate preferably 1–5 ml/min. The effluent from the column is directed to a RI detector or a UV detector based on the constituents of the feed to get a rough estimate of the elution of the components. For detailed analysis the effluent is also collected in 10 ml vials. The vials collected are then injected in a high-pressure liquid chromatograph to know the exact composition of the effluent stream. The results of the chromatograms are plotted on a graph to study the separation profile of the various components. **Figure 5** shows results of a pulse test conducted by the author to identify the elution profile of α-Tocotrienol and other feed components in palm oil extract feed.

9. Application of SMB in the industry

The first industrial scale SMB process was commercialized by UOP (United Oil Products check) for the separation of n-paraffins from branched paraffins, and aromatics. The first plant was started in 1960 and was used for the manufacture of biodegradable detergents. Traditionally p-xylene was purified using crystallization, however the SMB technology showed a big improvement over the crystallization process. After being successful in hydrocarbon separations there has been several novel processes that were developed for fatty chemicals, pharmaceuticals, carbohydrates, and biochemicals [10, 23–42]. Recently, a lot of process developments happened at Orochem Technologies with which the author was directly involved either in the R&D phase [22, 43–45] or in the commercialization phase [46–50].

9.1 Ion-exclusion SMB process

The extraction of sucrose from molasses have been researched extensively in the industry, in 1953, Dow Chemical company invented [51] an ion-exclusion process to separate the ionic and nonionic constituents of molasses. It was identified that under equilibrium conditions certain non-ionic components of molasses could be separated from the ionic components of molasses using SMB technology with high throughput and yield. This SMB technology can also be used for extracting sugar or related products from other biorefineries in addition to the one discussed in this section.

In 2012, the author was involved in developing a process using SMB technology to separate d-Tagatose from a mixture of d-Galactose and d-Tagatose produced in a biorefinery [43]. The process uses a strong acid cation exchange resin to provide a pure d-Tagatose product from a mixture of d-Tagatose, d-Galactose, glucose, and calcium salt. **Figure 6** shows a bench scale lab SMB that was run to find conditions to run the commercial scale SMB. In the lab scale setup, a 2–3-3 column configuration was established for the columns in a simulated SMB setup where a plethora of solenoid valves simulated the movement of the columns. Desorbent entered from top of 1st column and feed entered from top of 6th column. Extract was drawn from the bottom of 2nd column and raffinate was drawn from the bottom of 8th column. Column 1–2 are in desorption zone. Column 3–5 are in rectification zone and column 6–8 are in adsorption zone. After determining the flow conditions the flow rates were scaled up to run a commercial unit in Italy.

At industrial scale a continuous moving SMB was used for this process and a pilot unit was run to illustrate this process in Italy. The continuous SMB consisted of 30 ports to which 30 columns were connected. The 30 columns were separated into 3 sets of 10 columns each. **Figure 7** shows a schematic of the SMB process for separating d-Tagatose from d-Galactose using the continuous SMB. The desorbent which is water enters the 1st column as shown in **Figure 7**. The feed which is a mixture of d-Tagatose, d-Galactose, and salt enters from top of 5th column. Since the adsorbent has selectivity for d-Tagatose it gets preferentially adsorbed on the adsorbent and moves with it to come out from the bottom of 2nd column, this

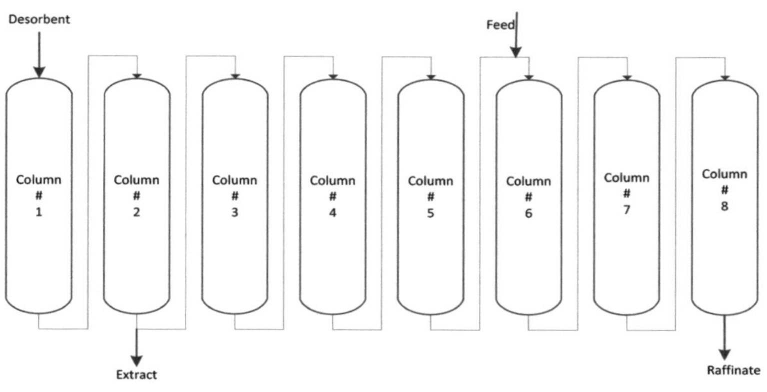

Figure 6.
Lab scale SMB process for separation of d-tagatose from d-galactose.

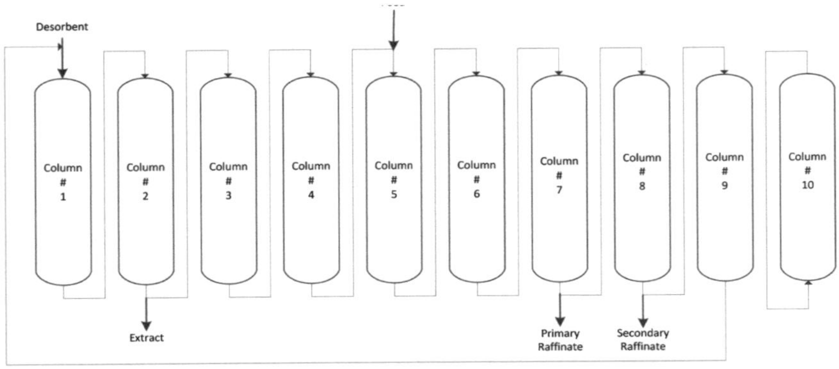

Figure 7.
Industrial scale SMB setup to separate d-tagatose from d-galactose.

stream is called extract and consists of pure d-Tagatose. D-Galactose the faster moving component comes out from the bottom of 7th column and is called the raffinate stream. In the continuous SMB a 2–3–2-1 column setup was employed to concentrate the raffinate stream and recycle part of desorbent making it a greener and more efficient process. Using SMB to concentrate streams saved considerably in energy cost by reducing the load on evaporators. This setup illustrates that not only can SMB be used to achieve an efficient separation with high purity and yield. The technology can also be used to build a greener process where process intensification can be used to integrate concentration steps to reduce the load on other unit operations such as evaporation.

9.2 SMB process to extract oxygenates from dilute fermentation stream

The author was involved in developing a novel process for extracting ethanol and other oxygenates produced in dilute fermentation processes in a biorefinery. Bio fermentation processes have extensively been used to produce products such as ethanol, non-condensable gases such as methane, oxygenated organic compounds such as 2,3 Butanediol (2,3-BDO). Oxygenated organic compounds have traditionally been produced from sugar sources such as corn, sugarcane, molasses, etc. The author has worked in collaboration with Lanzatech for isolating ethanol and other oxygenates produced in dilute fermentation broth. The process developed by Lanzatech uses unique microbes to convert carbon monoxide and other flue gases from power plant exhausts to useful chemicals such as ethanol, 2,3BDO. The Lanzatech process uses microbial gas fermentation to convert carbon monoxide containing gases produced by industries such as steel manufacturing, chemical production, and oil refineries as well as gases generated by gasification of forestry and agricultural residues, municipal waste, and coal into valuable fuel and chemical products to produce ethanol and other oxygenates such as 2,3-BDO [52, 53]. This process can also be used for extraction of products from other biorefineries working towards producing other products.

Since the ethanol and associated oxygenates such as isopropanol, 2,3-butanediol, and other diols are produced in very dilute concentrations in aqueous fermentation solutions, recovery of these constituents by traditional means such as distillation and crystallization is largely hindered by the energy requirements. The energy requirement for separating ethanol from a dilute mixture of ethanol and water is about 30,000 BTU/gal of ethanol produced. The author has developed a SMB process which can achieve the same separation with 75% less energy consumption [44].

9.2.1 Experiment for identifying capacity of adsorbent

A bench scale setup is used by the author to calculate the capacity of the adsorbent [44]. In the setup a column with a length of 250 mm and diameter of 10 mm is packed with the desired adsorbent. In this experiment the adsorbents chosen for comparison are activated carbon (E-325 Orochem adsorbent) and fluorinated activated carbon (E-325 an Orochem Technology proprietary product). A 1% solution of ethanol in demineralized water is passed through the columns with different adsorbents to find out the capacity of the adsorbents. The results of the adsorbent studies are shown in **Table 1**.

Another adsorbent studied for the SMB process was silica bonded with C-18. Since the performance of the fluorinated carbon was better than C-18 silica further analysis of the C-18 is not discussed in this chapter, more information can be found in the patent application [52].

9.2.2 Desorbent analysis

Once the adsorbent was finalized for the SMB process a solvent analysis was performed. The solvent selected should be selectively able to remove the adsorbed components without compromising the relative selectivity desired to achieve the separation. Solvents studied for this analysis were methanol, ethanol, propanol, methyl tertiary butyl ether (MTBE). In order to test the viability of the desorbent it was passed through a column saturated with the feed and eluted with the desorbent. 10 ml samples of the eluent was tested using a high-pressure liquid chromatogram to see if the desorbent was completely able to regenerate the column. Once regenerated the adsorption capacity of the adsorbent was calculated to understand if the regeneration was successful.

9.2.3 SMB setup

Lab scale SMB for extracting ethanol and butanediol from dilute fermentation broth was developed by the author. The SMB setup shown in **Figure 8** illustrates an 8-column setup where each column was packed with fluorinated carbon. The columns were connected in series to a bench scale SMB from Semba. Methanol was used as the desorbent. Synthetic feed solution containing 6% ethanol and 2% 2,3-butanediol was prepared and pumped to the top of column # 6. The complex SMB valve system facilitated switching of the inlet and outlet ports of the columns at regular intervals. The interval after which the inlet and outlet of the columns were switched is called the cycle time. The SMB setup is a continuous process where the desorbent and feed enters the columns and extract and raffinate are drawn out of the columns as designated in **Figure 8**. The switching of the SMB valve simulates a continuous counter current movement of the adsorbent and mobile phase. In the

	Exp#1	**Exp#2**	**Exp#3**	**Exp#4**
Organic @1% in water	Ethanol	2,3-Butanediol	Ethanol	2,3-Butanediol
Adsorbent	E-325	E-325	FC-5	FC-5
Breakthrough Volume (ml)	68	108	106	188
EtOH adsorbed(gm)	0.68	1.08	1.06	1.88
Adsorption ratio (W/W)	6.07	9.64	9.46	16.7

Table 1.
Below are the results of the adsorption capacity of the various adsorbents.

lab setup it was demonstrated that the extract stream did not have any water and the raffinate stream contained pure water. In order to achieve pure water at the raffinate stream the column had to be purged with nitrogen at 140°C to eliminate the methanol filled in the column. Since there was no way to establish the purging and heating of the column in the SMB setup the column was physically removed and replaced with a fresh column already purged with nitrogen at 140°C to simulate the concept. In the 2–3-3 SMB setup 2 columns between the desorbent and extract are in desorption zone, 3 columns between the extract and feed ports are in the rectification zone, and 3 columns between the feed and raffinate port are in the adsorption zone.

The lab scale setup explained in **Figure 8** was scaled up to as a 10 column SMB setup for industrial scale. **Figure 9** shows the 10 column SMB setup. The desorbent which was chosen to be ethanol as shown in **Figure 9** moves from left to right and the adsorbent beds switch from right to left simulating the counter current movement of the adsorbent and desorbent. During the next valve switch the column marked as column#1 moves to the position of column marked as column #10 and each column moves one position to the left. In the setup shown in **Figure 9** two columns are in an isolated regeneration mode. Column # 1 has steam passing

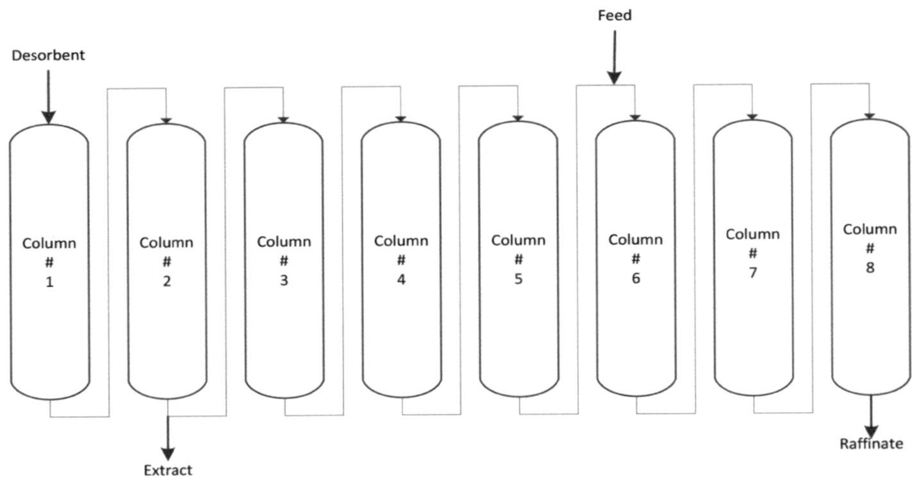

Figure 8.
Lab SMB setup for extracting ethanol and 2,3-BDO from dilute fermentation stream.

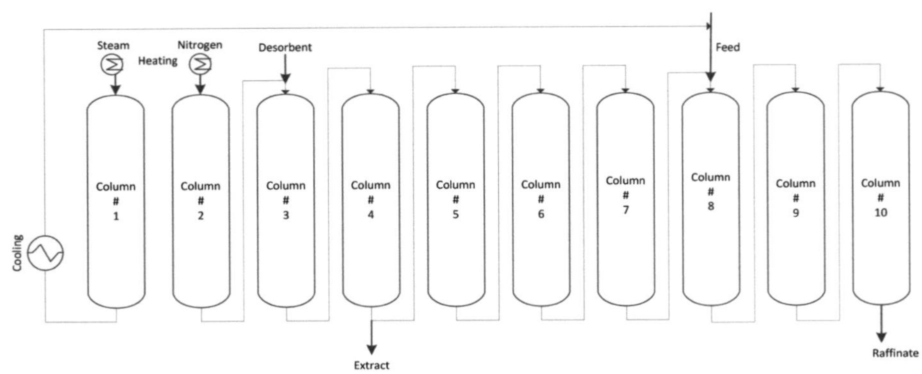

Figure 9.
Industrial setup to extract ethanol and 2,3-BDO from dilute fermentation stream.

through it at 100–120°C to eliminate any residues of ethanol – the ethanol water mixture derived from column # 1 is mixed with the feed entering the top of column # 8. Column # 2 has nitrogen heated to 80–120°C entering the column to strip as much ethanol as possible and mix with the desrobent which is ethanol. In the setup shown in **Figure 9** the raffinate contains only water which comes out from the bottom of column # 10. The 2,3-BDO contained in the feed comes out in the extract as a mixture of 2,3-BDO and ethanol.

The SMB processes explained in this section are an improvement over the traditional distillation or crystallization processes as the SMB process excessively reduces the amount of energy required to achieve higher purity and yield of the oxygenates produced in the fermentation process. Further optimization of the SMB process has been achieved by the author for several processes where the desorbent consumption can be reduced to achieve higher purity and yield of the feed components versus using any other separation technology.

In a further optimization of the SMB process for extracting ethanol and 2,3-BDO from dilute fermentation stream the author has developed a schematic depicted in **Figure 10**. In this SMB setup the same stationary phase can be used either fluorinated carbon or C-18 bonded silica [45]. For commercial scale a 15 column SMB setup is recommended where 5 columns can be dedicated to an improved and separate regeneration step from the rest of the 10 columns that perform the actual separation of the feed components. As depicted in **Figure 10** desorbent which is chosen to be ethanol is passed through the first column. Each subsequent column from 1 to 10 are connected in series like other SMB setup's such that the outlet of first column is connected to the inlet of the next column. A portion of the outlet stream from column # 3 is drawn as extract. The extract contains ethanol, 2,3-Butanediol and < 0.5% water. Feed enters the top of column # 7. Raffinate which is the water contained in the feed stream comes out of column # 9 and 10. In the improved SMB setup 5 columns are dedicated for improved regeneration. In the improved regeneration the columns are heated using superheated ethanol. As the columns step through the various zones once the column comes in the regeneration zone it is heated using superheated ethanol. The adsorbent bed is heated in 4 steps and eventually purged using nitrogen to remove the interstatial ethanol in the adsorbent bed. The columns are heated to about 110°C. Before the purged column must enter the adsorption zone it needs to be cooled to achieve the desired separation. The cooling is achieved by passing water coming out of the raffinate stream through the column to cool it to room temperature. The columns from 1 though 10 follow the same separation mechanism identified in **Figure 9**.

In order to heat the columns in the regeneration step superheated steam is passed through the columns, since the effluent from the columns would be hot it can be

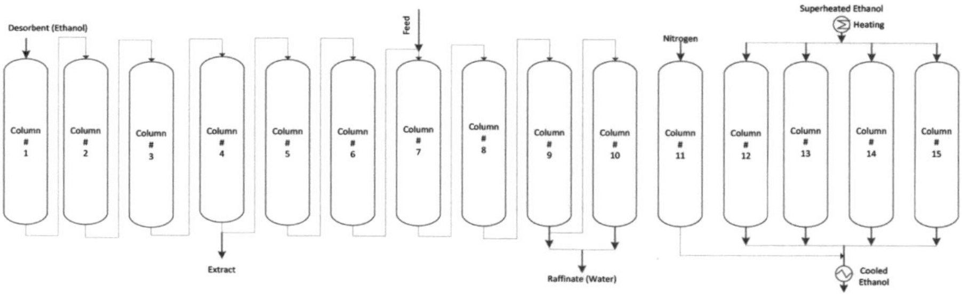

Figure 10.
Improved SMB setup for extracting oxygenates from fermentation broth.

cooled down using coolers and be used as the desorbent. The optimized SMB setup shown in **Figure 10** has removed the need to recycle a dilute ethanol stream from the SMB back to the feed stream as shown in **Figure 10**. If other desorbents like methanol, propanol, or MTBE were used for the separation then streams containing a mixture of desorbent and water would be created that would need to be distilled separately. More details on the different regeneration setups can be found in the work published by the author [44, 45].

9.3 Use of SMB for extracting biorefinery products

Most biorefineries produce products in dilute aqueous mixtures. For most cases the amount of energy required to separate the chemicals by traditional separation means is cost prohibitive, thereby making industrial scale production of biorefinery products impractical. The simulated moving bed technology provides the much-required unit operation for extracting biorefinery products at a fraction of the cost for traditional separation means. The author was involved in the extraction of fermentation products created in a bioreactor using SMB technology. If conventional distillation techniques were employed to separate the fermentation chemicals the technology would not have be feasible as the amount of energy required to separate the products using distillation was much more than the fuel value of the chemicals. The chemicals produced in the technology were ethanol and butanediol in a 6% and 2% concentration in the fermentation broth. Specifics of the technology can be found in Section 9.2. Until recently, most of the chemicals were produced by fossil fuel-based resources that are now depleting and are not sustainable. The depleting fossil fuels have catalyzed the use of biotechnology techniques for generating chemicals to meet the growing population. SMB technology has proven to be very successful in aqueous product streams. Moreover, the lower concentration of the product streams is favorable to SMB technology as it works on the principles of adsorption. In the examples outlined in this chapter SMB technology has demonstrated feasible for extraction of biorefinery chemicals such as ethanol, lactic acid, sugars (Tagatose, Galactose, Glucose), butanediol etc. As new processes are developed SMB technology can be applied to several biorefineries using techniques explained in this chapter to produce many other biorefinery products.

9.4 SMB in pharmaceutical industry

The use of SMB is not only limited to the petroleum, energy, biorefinery, or sugar industry. Recently there has been a lot of development in isolating active pharmaceutical ingredients using SMB technology. The author has developed a novel SMB process for isolating alpha-tocotrienol from palm oil extract [22]. In all the SMB processes we discussed the stationary phase used was non-polar and a polar solvent was used as desorbent. In the SMB setup for isolating α-tocotrienol [22] from palm oil extract the adsorbent used is either silica or alumina (polar adsorbent) and the mobile phase is a non-polar solvent. In this SMB setup α-tocotrienol need to be isolated from the rest of the components in palm oil extract like β-tocopherol/tocotrienol, μ-tocopherol/tocotrienol, δ-tocopherol/tocotrienol, front end and back end carotenoids. The SMB process developed creates a novel and green process due to the reduced amount of solvent required to achieve higher purity and yield of α-tocotrienol which would not have been possible using any other separation technology. The author has also contributed to the commercialization of the use of SMB technology for purification of highly pure EPA/ DHA from fish oil. The technology for purification of fish oil is documented in patents [46–50]. In addition to developing processes for achieving purification using SMB technology the author has worked on several other optimization projects for several other industries specifics of which can be found in publications referenced [54–56]. The author has also delivered several talks on safety topics with one of the topics on pump safety being published in CEP magazine [57]. The polar stationary phase SMB setup developed by the author can also be used to extract biorefinery products that are non-polar. Pharmaceutical products that are produced via fermentation can use SMB technology to achieve a cost effective separation with high throughput.

10. Future research directions

The use of SMB in biorefineries is gaining traction. With the growing population the need to conduct further research on using SMB for isolating biorefinery, pharmaceutical, and other biological products sustainably is imperative. The pharmaceutical, neurocritical, and biologics industry has primarily used batch processes which are limiting by capacity throughputs. The use of a continuous technology will create new opportunities in these industries and enable large scale production of products required to meet the growing demand of the world.

11. Conclusion

Adsorption processes have successfully been demonstrated for purification of commercial products. For those processes SMB technology has shown great advantage over batch chromatography. One can achieve higher yield and purity using SMB technology. Various process developments in using SMB for multi component feed streams have been successfully commercialized. Process Intensification can be achieved using this novel technology in biorefineries. Fermentation products that are produced in dilute streams have found an efficient way of extraction using the SMB technology. Higher throughput processes in the pharmaceutical and bio-fermentation industry will enable the use of SMB technology. The successful implementation of SMB will enable increased capital investment for R&D related to this technology.

Several processes have been developed using SMB technology to achieve separation in biorefineries: e.g. purification of glycerol from biodiesel production (using the Ambersep BD50 resin, or gel-type acidic ion-exchange resin beads) where the raffinate stream contains salts and organic impurities including free fatty acids, purification of oligosaccharides (made up of xylose and arabinose units), isolation of lactic acid from acetic acid, separation of sugars (glucose and xylose) [58] to name a few. The SMB technology has shown immense promise in achieving a sustainable, cost effective, and safe product isolation technology for various industries with equivalent promise for biorefineries. Future research in the field has a bright future for developing industries such as biorefineries.

Acknowledgements

First and foremost, I would like to acknowledge almighty god for giving me the opportunity and strength to be a contributor. My sincere gratitude to my parents, my family, and specially my cousin Sujata Sharma for always believing in me and being an inspiration and support structure.

I would also like to acknowledge Orochem Technologies for giving me the opportunity to work on cutting edge SMB technology and for enabling me to be part of the development of these technologies.

Author details

Deepak Sharma
Bayer U.S. LLC, Engineering and Technology, St. Louis, Missouri, USA

*Address all correspondence to: deepak.sharma2@bayer.com; deepaksharma@wustl.edu

References

[1] S. Eagle and J.W. Scott. Refining by adsorption: New cyclic adsorption process separates high purity aromatics and olefins from petroleum. Petrol. Process., Aug., 881-884 (1949).

[2] C.M. Hansen, Hansen Solubility Parameters: A User's Handbook, CRC Press, Boca Raton, 2007.

[3] S.E. Koonin, Getting serious about biofuels, Science, 311, 435-435 (2006)

[4] R.C. Binning and F.E. James, Permeation. A new commercial separation tool, Perot. Eng., 30, 6 (1958)

[5] T.C. Ezeji, N. Qureshi and H.P. Blaschek, Butanol fermentation research: Upstream and downstream manipu-lations, Chem. Rec., 4, 305-314 (2004).

[6] M.E. van Leeuwen, Derivation of Stockmayer potential parameters for polar fluids, Fluid Phase Equlibr., 99,1-18 (1994)

[7] C. Reichardt, Solvents and Solvent Effects in Organic Chemistry, Wiley-VCH Verlag GmbH, Weinheim, 1988.

[8] C.M. Grill, L. Miller and T.Q. Yan. Resolution of a racemic pharmaceutical intermediate: A comparison of preparative HPLC, steady state recycling and simulated moving bed. J. Chromatogr. A, 1026, 101-108 (2004).

[9] Chem. Eng. 39, (Aug. 29, 1977)

[10] Stanley A. Gembicki, James Rekoske, Anil Oroskar, James Johnson Kirk-Othmer Encyclopedia of Chemical Technology, https://doi.org/10.1002/0471238961.0104191507051302.a01.pub2

[11] U.S.Pat. 4,434,051, (February 28, 1984), M. W. Golem (To UOP)

[12] U.S.Pat. 5,565,104, (October 15, 1996), J. W. Priegnitz (To UOP)

[13] E. Herschler and T. S. Mertes, Ind. Eng. Chem. 47, 193 (1955).

[14] D. Haresnape, F. A. Fidler, and R. A. Lowry, Ind. Eng. Chem. 41, 2691 (1949).

[15] J. Mair and M. Shamaiengar, Anal. Chem. 30, 276 (Feb. 1958).

[16] J. Mair, A. L. Gaboriault, and F. D. Rossini, Ind. Eng. Chem. 39, 1072 (1947).

[17] S. Eagle and J. W. Scott, Ind. Eng. Chem. 42, 1287 (1950).

[18] E. Hirschler and S. Amon, Ind. Eng. Chem. 30, 276 (Feb. 1958).

[19] Heftmann, ed., Chromatography, Van Nostrand-Reinhold, New York, 1975.

[20] J. J. Kipling, Adsorption from Solutions of Non-Electrolytes, Academic Press, Inc., New York, 1965

[21] C. Nachod and J. Schubert, Ion-Exchange Technology, Academic Press, Inc., New York, 1956.

[22] U.S. Pat. 2014/0179933 A1, (26 June, 2014), Oroskar et. al., (To Orochem Technologies)

[23] U.S.Pat. 6,222,088, (April 24, 2001), S.Kulprathipanja, (To UOP LLC)

[24] U.S.Pat. 5,223,589, (June 29, 1993), S.Kulprathipanja, (To UOP LLC)

[25] U.S.Pat. 5,177,300, (January 5, 1993), S.Kulprathipanja, K.K.Kuhlne, M.S.Patton, R.L.Fergin)To UOP LLC)

[26] U.S.Pat. 5,159,131, (October 27, 1992), H.A.Zinnen, (To UOP LLC)

[27] U.S.Pat. 5,143,586, (September 1, 1992), B.McCulloch, (To UOP LLC)

[28] U.S.Pat. 5,019,271, (May 28, 1991), H.A.Zinnen (To UOP LLC)

[29] U.S.Pat. 5,004,853, (April 2, 1991), P.T.Barger, T.J.Barder, D.Y.Lin, S.H.Hobbs, (To UOP LLC)

[30] U.S.Pat. 4,992,618, (February 12, 1991), S.Kulprathipanja

[31] U.S.Pat. 4,876,390, (October 24, 1989), B.McCulloch, (To UOP LLC)

[32] U.S.Pat. 4,784,807, (November 5, 1988), H.A.Zinnen, (To UOP LLC)

[33] U.S.Pat. 5,276,246, (January 4, 1994), B.McCulloch, J.R.Lansbarkis, (To UOP LLC)

[34] U.S.Pat. 5,220,102, (June 15, 1993), G.A.Funk, J.R.Lansbarkis, A.R.Oroskar, B.Mcculloch (To UOP LLC)

[35] U.S.Pat. 5,177,295, (January 5, 1993), A.R.Oroskar, R.E.Prada, J.A.Johnson, G.C.Anderson, H.A.Zinnen, (To UOP LLC)

[36] U.S.Pat. 5,149,887, (September 22, 1992), H.A.Zinnen, (To UOP LLC)

[37] U.S.Pat. 5,071,560, (December 10, 1991), B.McCulloch, W.H.Goodman (To UOP LLC)

[38] U.S.Pat. 5,012,039, (April 30, 1991), T.J.Barder, (To UOP LLC)

[39] U.S.Pat. 4,992,621, (February 26, 1991), B.McCulloch, M.G.Gatter, (To UOP LLC)

[40] U.S.Pat. 4,977,243, (December 11, 1990), T.J.Barder, B.W.Bedwell, S.P.Johnson, (To UOP LLC)

[41] U.S.Pat. 4,797,233, (January 10, 1989, H.A.Zinnen, (To UOP LLC)

[42] U.S.Pat. 4,770,819, (September 13, 1988), H.A.Zinnen, (To UOP LLC)

[43] U.S.Pat. 8,802,843 B2, (12 August 2014) Oroskar et al., (To Orochem Technologies LLC)

[44] U.S.Pat. 8,658,845 B2, (25 February, 2014), Oroskar et al., (To Orochem Technologies LLC)

[45] U.S.Pat. 8,704,016 B2, (22 April, 2014), Sharma et al., (To Orochem Technologies LLC)

[46] U.S. Pat. 9,790,162, (17 October, 2017) Adam et al., (To BASF Pharma (Callanish) Limited)

[47] U.S. Pat. 9,771,542, (26 September, 2017) Adam et al., (To BASF Pharma (Callanish) Limited)

[48] U.S. Pat. 9,695,382, (04 July, 2017) Adam et al., (To BASF Pharma (Callanish) Limited)

[49] U.S. Pat. 9,493,392, (15 November, 2016) Adam et al., (To BASF Pharma (Callanish) Limited)

[50] U.S. Pat. 9,370,730, (21 June, 2016) Adam et al., (To BASF Pharma (Callanish) Limited).

[51] M. Wheaton and W.C.Bauman, I&EC Eng. And Proc. Dev 45, 228 (1953)

[52] U.S. Pat. 20100323417A1 Carbon capture in fermentation, 2007

[53] U.S. Pat. 8,119,378 Microbial alcohol production process

[54] Karre, A.V.; Sharma, D.; Valsaraj, K.T. Estimating fouling and hydraulic debottlenecking of a clarifier piping system in the expansion of a chemical manufacturing plant. Chem. Prod. Process Modeling 2020, 20200029, doi:10.1515/cppm-2020-0029.

[55] Sharma, D.; Karre, A.; Valsaraj, K. Evaluations of the capacity of an existing brine system and estimation of salt loading profile for increased soft water demand to avoid soil contamination. Can. J. Chem. Eng. 2020, doi:10.1002/cjce.23937.

[56] Sharma, D.; Karre, A.V.; Valsaraj, K.T.; Sharma, S. Intensification of a Neutralization Process for Waste Generated from Ion Exchange Regeneration for Expansion of a Chemical Manufacturing Facility. Processes 2021, 9, 1285. https:// doi.org/10.3390/pr9081285

[57] Sharma, D. Ensure safe pump operation, CEP Magazine, Feb 2021

[58] Kiss, A. A., Lange, J., Schuur, B., Brilman, D. W. F., van der Ham, A. G. J, Kersten, S. R. A., Separation Technology – Making a difference in biorefineries. Biomass and Bioenergy May 2016, DOI:10.1016/j.biombioe.2016.05.021

Chapter 3

Enzymes – Key Elements of the Future Biorefineries

Klara Birikh, Alex Michine, Matti Heikkilä and Petri Ihalainen

Abstract

The biorefinery concept in its modern meaning has emerged after it has become apparent that biofuel production from non-food biomass is struggling for economic viability. Lignocellulosic biomass is more recalcitrant and more complex than the starch-based feedstocks used for food. The former, therefore, calls for a more complex approach to its utilization. This chapter reflects MetGen's vision of the future development of biorefineries. We will discuss the zero-waste approach to lignocellulosic biomass utilization and various ways to valorize the resulting streams to boost the economic viability of the biorefinery. We will mostly explore the relevant enzyme-based approaches and will make a special focus on lignin valorization. Enzymatic and cell-based approaches to sugar valorization will be discussed as well.

Keywords: enzyme, lignocellulosic biomass, lignin, cellulose, hemicellulose, laccase

1. Introduction: from bioethanol to biorefineries

The progenitor of the modern biorefinery concept was bioethanol production. In the 1970s, Brazil and the United States started mass production of bioethanol grown from sugarcane and corn respectively. The most common usage of bioethanol is to power automobiles by mixing it with petrol. The sugar yield from these feedstocks is very high and the biomass processing is rather simple, thus fueling the transportation this way was economically viable, especially in the countries with scarce fossil fuel resources, like Brasil.

Bioethanol is generally CO_2 neutral because the released during the burning of ethanol is compensated by the absorption of the CO_2 by growing the feedstock biomass. This however does not consider the CO_2 generated by the logistics of the biomass production and processing. Besides, blending bioethanol with gasoline helps to reduce greenhouse gases (GHG) emissions by oxygenating the fuel mixture which makes it burn more completely. Thus bioethanol was considered to be an environmentally friendly alternative to petrol.

In the future, with improved efficiency, utilization of non-agricultural feedstocks and use of renewable energy, the respective life cycle GHG emissions could be cut by up to 86 percent relative to gasoline as reported in EPA's Emission Facts [EPA (2007) Emission Facts; Greenhouse Gas Impacts of Expanded Renewable and Alternative Fuels Use. Emission Facts Report (EPA420-F-07-035). Office of Transportation and Air Quality, EPA, US].

Thus the agenda of bioethanol production was shifted to the products derived from lignocellulosic biomass to avoid competition with food and limit the use of agricultural land.

In the brink of the 21st century a considerable public and private effort to implement the so-called second-generation bioethanol industry based on lignocellulosic, non-edible feedstock was undertaken, and eventually faded away due to economic inefficiency. The frustrating experience of lignocellulosic bioethanol hype of the past years triggered the formation of a broader view of the biorefinery concept. It grew with the understanding that if only a part of the biomass, namely the cellulose, is used to make a product, moreover, not a high-value product such as bioethanol, the economics of such an undertaking does not work [1].

This notion coincided with the growing understanding that biomass is not merely a quick fix for a deficit of fossil resources in some countries, but a fundamental raw material for bioeconomy. Consequently, a biorefinery concept was forming with the term borrowed from the petroleum oil refinery, which goes beyond the exhaustion of biomass into a spectrum of products. Biorefineries are based on four principles [2], namely principles of sustainability, cascading, non-conflict with food, and neutral carbon footprint.

Thus to implement the biorefineries as fundamental units of bioeconomy, all biomass components cellulose, hemicellulose, and lignin need to be utilized. Moreover, the general approach should be similar to the oil refinery concept - the raw material needs to be fractionated to result in a range of intermediates leading to a variety of products from high to low-value. Ideally, there has to be in-built flexibility allowing to change the product portfolio according to the current market demands.

One of the fundamental differences of biorefineries from oil refineries is the repertoire of tools that can be used, where enzymes - the natural catalysts play an important role.

Nature is using biocatalysts – the protein molecules called enzymes - for performing virtually all biochemical reactions happening inside organisms, and often outside as well.

It is logical to assume that a bio-based economy would be largely relying on biocatalysis. Extremely high specificity and selectivity as well eco-friendliness make enzymes potentially very attractive in many industrial applications.

2. Biorefinery as an industry sector

Following the concept of Biorefineries, the society has to make a leap from biofuel factories using local agricultural feedstocks to produce bioethanol for fulfilling local demand for automobile fuel to the biorefineries providing raw materials for various industries from energy to chemicals and materials producers. From this perspective, wood seems to be the most likely feedstock to be able to fulfill the industry demand.

It has to be noted that biomass alone, wood or other types, is unlikely to fulfill the energy demand of modern society from the volume point of view. Other energy-providing technologies, like solar and wind energy need to fill in the gap. However, biomass is suited to assume other roles in a circular economy, related to materials and chemicals. This notion provides only more motivation for diversifying the biorefineries' product range.

Wood is one of the most abundant, sustainable raw materials on Earth, which is available around the year. It requires no roof for storage and has a high density, which is favorable for logistics and handling. Furthermore, it requires no additional field space and has no agricultural or nutritional use.

If wood is the most likely feedstock of the rising bioeconomy, then the pulp and paper industry is the most likely first block in the value chains of the biobased products. This industry has many years of experience in maintaining and working forests, as well as harvesting, transporting, and processing wood. It is also noteworthy that in the Nordic countries, the volume of sustainably harvested forests is growing faster than the current consumption: regulation and standardized systems are in place to allow forests to be harvested sustainably to meet significant industrial demand.

The drawback of this industry being at the foundation of the biobased economy is that this industry has highly refined processes focusing on cellulose fibers only, the industry is highly conservative due to low-margin economic positioning and besides this industry in its current state is used to offering a very narrow product portfolio which is marketed through distributors.

Diversifying the product offering of the pulp and paper industry may require changes in the processes or the addition of parallel process lines and more intimate interaction with various markets. This in turn requires investments and a change in attitude.

Biorefineries can be positioned on the interface of pulp and paper/forestry industry and chemicals and materials industry. And it has to be admitted that this interface is yet to be created. For example, it can be implemented with a third party operating a biorefinery with the over-the-fence supply of raw materials (feedstocks) and possibly even utilities from a pulp and paper mill. These feedstocks may comprise lignin, zero fibers (short fibers disposed of with the wastewater), pulp products depending on the demand of both markets. This could be set up as a joint venture so that both organizations can benefit from this model. Alternatively, a joint venture with the end-user of the biorefinery products can be envisaged as well.

Enzymes as an important part of the economics and the technology of the biorefineries can also be considered as part of the production process. On-site manufacturing of enzymes allows saving costs on concentration, formulation, storage, and shipping of the enzymes. Some companies embracing biorefineries, develop their own enzymatic solutions to be implemented in their biorefineries, and can set up on-site manufacturing at their will. Whereas other types of biorefinery owners, like pulp and paper companies, usually rely on an external enzyme supplier. Usually, enzyme suppliers are not open to providing their production strains to third parties for on-site manufacturing. In this respect, MetGen has a more flexible business model towards supplying enzymatic solutions for biorefineries, including a possibility of on-site manufacturing.

3. Enzymes—ultimate tools for biobased industries

Many if not most industrial chemical processes are dependent on catalysts - substances that accelerate chemical reactions without themselves being consumed in the catalyzed reaction and can continue to act repeatedly. Because of this, only very small amounts of catalyst are required to have a dramatic effect on the reaction rate. The development of affordable durable and efficient catalysts was vital for the establishment and economic viability of fossil-based chemistry and material science.

It is equally important for the biobased economy to adapt and further develop nature's catalytic tools.

The historic concern about enzymes is that they are vulnerable to industrial conditions and often could not be applied to existing industrial processes. Modern molecular biology and bioengineering pave the way to much wider use of enzymes in the industry by making it possible to adapt enzymes to performing in unnatural harsh conditions. The development time for new enzymes was further reduced with the development of bioinformatics tools and genome editing.

Especially as new bio-based processes are being developed it is a good time to consider making them more enzyme-adaptable by assuming somewhat longer retention times while transitioning to lower temperatures and pressures, as compared to currently common conditions.

Enzymatic processes are truly similar to chemical catalysis. They can be run as homogeneous catalysis with a soluble enzyme added and disposed of with every production batch, or as heterogeneous catalysis, where the enzyme is used in the immobilized form and reused from batch to batch or used in a continuous process with a column set up.

Importantly, the enzymes present also a third option not applicable with the chemical catalysts - a continuous membrane bioreactor. Sometimes this technology is called "enzymes immobilized by perfusion". This setup exploits the best of the previous two - affordability of the soluble enzyme and reusability of the immobilized one. In this setup, the enzyme is trapped in a bioreactor connected to a tangential flow micro-filtration membrane unit allowing the low-molecular-weight product to penetrate through the membrane but retaining the enzyme inside.

This setup allows not only an efficient use of the enzyme but can also provide product fractionation and more complex designs with parallel processes. Ultra and nano-filtration is also a very useful and economical water removal tool. We will further discuss this setup in the section dedicated to lignin valorization.

3.1 Bioconversion—enzyme or whole cell?

One important aspect of enzyme-dependent catalysis is the necessity of a cofactor for some enzymes. Cofactors are important accessories to biochemical processes. They are small organic compounds or metal ions empowering enzymes to function at maximal catalytic effectiveness or endurance. Cofactors may aid in substrate binding, catalysis, stabilizing the transition state, or contributing to the overall stability of the enzyme's structure. In some cases cofactors are modified during the reaction, for example, providing or accepting an electron in reduction–oxidation reactions, or providing energy through a high energy bond braking. In this case, in order to be reused, cofactors need to be regenerated during the reaction - oxidized/reduced/phosphorylated respectively. Regeneration requires another enzyme and a co-substrate to be oxidized or reduced. With the cofactor regeneration in place, the reaction can proceed continuously with only a small amount of the cofactor present. The chemistry of cofactor regeneration is well known nowadays [3]. The challenge is mostly regarding how to achieve the regeneration with immobilized enzyme systems which are preferred for industrial processes to facilitate the recovery and continuous use of the catalysts. This has become a great hurdle for the industrialization of many promising enzymatic processes. Once again, recent advances in membrane technologies led to the development of sustainable methods based on membrane entrapment [4].

Nevertheless, the necessity of a cofactor complicates the enzymatic process and increases the cost. Thus most of the bio-transformations involving cofactors have been traditionally performed in the industry with living cells often referred to as microbial cell factories [5].

Whole-cell biotransformation has advantages and disadvantages as compared to the enzymatic process. As mentioned before it solves the problem with the cofactors as they are widely used in cell metabolism and regeneration routes are in place. The balance of cellular metabolic fluxes can be further genetically adjusted for the increased level of the components necessary for the product synthesis. Another advantage of the microbial cell factories is that multistep reactions can be carried out, and it is often possible to use simple and affordable raw materials such as glucose because the cell has a metabolic pathway in place to convert it to a large variety of precursor and eventually to the final product. Among the shortcomings of the cell factories, one should mention a very narrow operational space due to microorganisms viability constraints. While some individual enzymes can tolerate high temperatures close to water boiling point and a wide range of pH, industrial microorganisms are usually performing only in ambient conditions. Besides, there are often multiple pathways in the cell to convert the starting material, which leads to the formation of side products. In more detail, the cons and pros of enzymatic and whole-cell bioconversions are listed in **Table 1**. In conclusion, as opposed to whole microorganism bio-conversions, more common in the past, enzymes provide faster and safer processes with a broader operational range.

Enzymes are also attractive in industrial use from the safety point of view: enzymes are not living organisms and they cannot breed (as opposed to the whole-cell factories) and can be considered environmentally safe. Additionally, being proteins, enzymes do not create toxic waste and decompose naturally over time. It should be noted that enzymes – as is the case with all proteins – may cause allergenic irritation. Therefore, the use of highly concentrated industrial enzymes should always be done according to handling instructions and material safety documentation.

Thus, with all the pros and cons in mind, the preferred type of bioconversion needs to be identified for each particular process.

Respective sectors of molecular biology dealing with enzymatic processes and whole-cell bioconversions are Enzyme engineering [6] and Metabolic engineering [7] respectively. Where enzyme engineering provides tools for optimizing protein structure for better performance; and metabolic engineering provides tools for optimizing the microbial genome to redistribute metabolic pathways in favor of the desired product formation. It has to be noticed that industrial process engineering is extremely important to go hand in hand with molecular engineering.

3.2 Enzyme resources of nature

Enzymes are extremely abundant in nature and exist in all living organisms from bacteria to humans. All industrial enzymes have their origin and prototypes in nature, where wood is decomposed by rot microorganisms, the most efficient

	Enzymatic process	**Whole call process**
Temperature, pH range	Wide	Narrow
Substrate/product load	High	Low
Tolerance to solvents	Moderate	Low
External cofactors/cofactor regeneration system	Needed	Not needed
Multistep processes	Difficult	Natural
Control over the reaction speed	By increasing the enzyme concentration	Only by increasing the size of the vessel
Side products	No	Yes

Table 1.
Enzymatic process vs. whole-cell bioconversion.

of which are fungi. It is thus natural that most enzymes used in industry for wood and other biomass applications are of fungal origin. Robust industrial strains and processes for fungal enzyme production have been developed through decades of optimization.

One of the major problems of fungal enzymes is that fungi are not known to live in extreme environments, such as elevated temperatures and extreme pHs, and their enzymes are usually not tolerant to harsh industrial conditions, which is sometimes limiting their application. In contrast, bacteria populate such environments as hot springs, salt lakes, and ocean depths. Some bacteria also possess individual enzymes with relevant catalytic activities for industrial biomass applications. Recent advances in molecular biology, genome sequencing, and genetic engineering made bacterial enzymes an attractive alternative to their fungal counterparts. Bacteria offer more diverse natural prototypes, and there are better-developed tools for genetic engineering in bacteria, allowing further optimization of the enzymes to required conditions.

Some unique industrial enzymes of bacterial origin have been developed in the past years breaking the boundaries of industrial enzyme applications. Nevertheless, when multiple enzymes are required in one process, such as cellulases, fungal production is usually a preferred option.

4. Biomass hydrolysis

4.1 Biomass pretreatment

The biorefinery platform requires pretreatment of lignocellulosic materials, which can be very recalcitrant, to improve further processing through enzymatic hydrolysis, and for other downstream unit operations.

Pretreatment employs a combination of chemical and physical elements such as temperature, pressure, and acid or alkali. This partially separates biomass components such as cellulose, hemicellulose, and lignin from each other resulting in a paste-like rather than a solid substance. This level of destruction allows access of enzymes to all the biomass components and further separation and hydrolysis.

Many pretreatment methods and unit operations were inherited from the bioethanol-oriented processes, where the target product was a fermentable sugar mix, and the ultimate goal to reduce the cost of the process. Now, when the focus of the biorefinery concept has shifted from the design of more or less energy-driven biorefineries to much more versatile facilities where chemicals and other raw materials can be produced apart from energy carriers, the view to the pretreatment has been transforming as well. In some cases, a pretreatment with a higher cost, but also better separation of the biomass components and higher quality streams are preferred. For example, organosol or chemical treatment employing ionic liquids and deep eutectic solvents. In the end, the choice of pretreatment must be based on a thorough techno-economic evaluation considering the proposed applications and the source of the biomass. This topic is reviewed in detail elsewhere [8, 9].

4.2 Cellulases for biomass hydrolysis

It is well known that wood is efficiently decomposed in nature by filamentous fungi. In their natural habitat, these microorganisms live on a solid substrate like wood and secrete a number of hydrolytic enzymes degrading all wood components down to low molecular weight substances that can be used as nutrients. In industry, the enzyme preparations were traditionally obtained by the propagation of the fungal strains in a liquid medium, and such production method resulted in a cocktail of different enzymatic activities, often generally referred to as cellulase [10]. Fungal metabolism has a complex regulation in order to be able to produce the set of enzymes relevant to the available type of biomass [11]. Thus, for example, cellulase production by the fungal cells is induced by certain compounds generated in wood hydrolysis. The enzymatic cocktail produced by a fungus depends on the fungal strain properties and is not always optimal for a particular industrial application.

The most noticeable hurdle for the industrial application of natural fungal cellulase cocktails is the mechanism preventing glucose accumulation in the environment of the fungal cell. Such accumulation could provide a favorable environment for competing microbes such as bacteria, which cannot degrade wood themselves. To achieve this regulation, all of the enzymes of the cellulase cocktail are inhibited by their reaction products [12]. As seen in **Figure 1**, cellulose is initially attacked by a number of enzymes most prominent one - exoglucanase (also known as cellobiohydrolase or CBH) comprising more than 50% of total protein in the cocktail, which is assisted by accessory enzymes endoglucanase, oxidative cellulase lytic polysaccharide monooxygenase (LPMO), and indirectly by other enzymes. The concerted action of these enzymes results in the formation of glucose dimer, cellobiose. This is followed by the last step of the cellulose hydrolysis, splitting cellobiose to two glucose, performed by beta-glucosidase. Glucose is further absorbed by the cell and metabolized. If the hydrolysis proceeds faster than glucose is consumed causing glucose accumulation, beta-glucosidase is inhibited by glucose and slows down, this, in turn, results in cellobiose accumulation slowing down exoglucanase, and thus the entire chain of the reactions is regulated by the feedback response from the last step (as shown with red arrows in **Figure 1**). However, in industrial biomass hydrolysis, glucose accumulation is the ultimate goal. Thus, this feedback loop needs to be overruled. This is usually done by artificially increasing the amount of beta-glucosidase in the cocktail. This can be done by inserting additional genes for beta-glucosidase into the fungal strain. Besides, some beta-glucosidases are less inhibited by glucose (or more glucose-tolerant) than others, and this can also be exploited in composing industrial cellulase cocktails. It has to be noted that glucose tolerance of beta-glucosidases is poorly understood and occurs more often in bacterial enzymes than in fungal ones. Elucidating the molecular mechanisms of glucose tolerance is a very important aspect of cellulose biotechnology research of glucose tolerance [13].

Apart from providing a feedback regulation of hydrolysis speed, beta-glucosidase has another important function - generating the inducers of cellulase gene expression and ultimately the cellulase production. Those inducers are unusual glucose dimers of which sophorose, a glucose dimer with β-1,2 bond, is the most efficient one. In nature, these compounds appear in small amounts in the presence of cellobiose as a result of a side-activity of some beta-glucosidases. This activity is referred to as trans-glycosylation [14]. The molecular mechanism underlying the ability of the beta-glucosidase to perform trans-glycosylation is obscure. There are

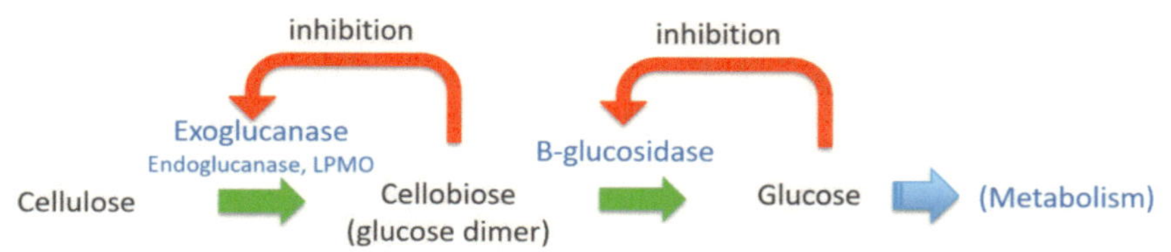

Figure 1.
Natural regulation of cellulose hydrolysis.

several different glucosidases in the fungal cell, extracellular as well as intracellular. Most probably it is the intra-cellular beta-glucosidases that are responsible for the inducer generation in nature [15].

In the industrial hydrolysis process, the inducer sugar is produced artificially from glucose either using random chemical dimerization catalyzed by phosphoric acid under elevated temperature and pressure or using beta-glucosidase with transglucosylase activity.

Thus in cellulose hydrolysis, CBH is the most prominent enzyme comprising more than a half of the total protein content of the cocktail, however, beta-glucanase is the most important enzymatic tool to providing cellulase efficiency in industrial hydrolysis applications.

Biochemists and molecular biologists have been studying the components of fungal cellulases and identified specific proteins responsible for the degradation of various plant polymers such as cellulose, other glucans, pectin, xylan, mannan, lignin, etc.

The native cocktails were improved after discovering that certain additional activities could enhance the conversion rates of specific biomass feedstock types [16–19]. For example, feedstocks could comprise different plants, pretreatments, or combinations of both. These augmented cellulases were produced by blending different secretomes containing the desired activities. More recently, some required activities have been genetically engineered into production strains.

Despite the considerable improvements in general-purpose cellulases available from the main enzyme-producing companies, a "one-size-fits-all" cellulase does not effectively address the wide range of biomass type-pretreatment chemistry combinations. However, customization of the cellulase cocktail is not commonly offered by enzyme producers. On the contrary, MetGen offers customization of the hydrolysis solution MetZyme® SUNO™ to the client's specific substrate/pretreatment as well as an option for on-site enzyme manufacturing.

4.3 Special enzymes for biomass degradation and valorization

Biomass is mostly comprised of polymers. The main structural components of plant biomass are polysaccharides (cellulose, hemicellulose, starch, pectin, and other plant polymers), and polyphenols (lignin). All these polymers apart from cellulose are branched and diverse in structure. In the plant cell, they are interlaced to form a complex and often recalcitrant structure.

Natural fungal cocktails are instrumental in the full hydrolysis of the biomass to monomers or low molecular weight compounds. This was the mainstream strategy of non-food biomass processing in the biofuel era. When we think of wider and wiser use of the biomass in a range of industrial applications it may appear sensible to preserve the polymeric structure of certain components. This can be partly achieved by choosing the right pretreatment method and other chemical and physical methods of processing the streams. Further, the process of biomass fractionation can be tuned by enzymes.

Generally speaking, the same enzyme toolbox is applicable both in the biorefinery industry and in Pulp and Paper sector. Notably, however, while in the P and P industry enzymes are minor and optional components of the process, in the biorefinery concept, enzymatic processes play a major role and represent a major cost, thus also opening a major market.

Let us consider how various product streams from biomass can be tuned by specific enzymes. An overview of the individual enzyme activities used in enzymatic solutions provided by MetGen is given in **Table 2**.

Generic activity	Application	Reaction conditions	MetZyme® family
Cellulase (cocktail)	Biomass hydrolysis	55 C, pH 5.0	MetZyme® SUNO™
Endoglucanase	Cellulose fiber modification, nanocellulose production; biomass hydrolysis;	High/ambient temperature, acidic/alkaline pH,	MetZyme® BRILA™ components
Xylanase	Hydrolysis of xylan for fiber modification; customization of hydrolysis cocktail	High/medium temperature, acidic/alkaline pH	MetZyme® BRILA™ optional components MetZyme® Suno optional components
Mannanase	Hydrolysis of mannan for fiber modification; customization of hydrolysis cocktail	High/ medium temperature, acidic pH	MetZyme® BRILA™ optional components MetZyme® SUNO™ optional components
Pectinase	Hydrolysis of pectin for fiber modification; customization of hydrolysis cocktail; R&D – pectin valorization	High/ medium temperature, acidic pH/highly alkaline pH	MetZyme® BRILA™ optional components MetZyme® SUNO™ optional components
Glucose isomerase	Low purity glucose to fructose conversion for platform chemicals production (especially suited for lignocellulosic glucose).	High temperature Acidic to moderately alkaline pH	MetZyme® PURECO™-GI
Pyranose oxidase	Glucosone from Glucose	Medium temperature Acidic to moderately alkaline pH	MetZyme® PURECO™-Pyranose oxidase
Lytic polysaccharide monoxygenase (LPMO)	Fiber modification (adding charge to cellulose fibers)		Not commercially available. Lab samples are available from MetGen for joint product development.
Laccase	Phenol/Polyphenol oxidation		METNIN™ – lignin refining technology MetZyme® BRILA™ optional components MetZyme® SUNO™ optional components

Table 2.
MetGen's product families and respective enzymatic activities.

4.3.1 Lignin valorization

In the earlier biorefinery concepts, lignin was often mostly regarded as a recalcitrance factor, fermentation inhibitor, sugar stream contaminant, etc. A broader view of the biorefinery, however, considers the valorization of lignin as a vital component of the economics of the entire concept. This is why it is one of the fastest-growing research and development areas in the biomass valorization field [20–23].

The main hurdles of lignin valorization are its diverse structure and poor solubility. Liquefaction of lignin would allow its use as fuel, as it is reached in high-energy chemical structures. More precise depolymerization or fragmentation of lignin may enable higher-value products for various industries from construction to high-performance materials. Even though lignin-based replacement products have already been reported [20, 24–26] to be useful as binders, coatings, and fillers, and others, these applications are not yet widely industrially implemented. The main challenges for the full valorization of lignin are the economical production of suitable lignin and maintaining consistent quality throughout different batches. In order to achieve desirable properties for the industrial application, lignin usually needs to be fragmented and refined to a lower molecular weight and often chemically modified as well.

Enzymatic degradation of lignin, which occurs in nature, was speculated for a long time to be applicable in the industry [27]. This approach seems attractive because the catalysis takes place in water and under mild conditions avoiding high pressure, temperatures, and hazardous and expensive chemical catalysts, thus saving CAPEX and lowering environmental impact.

Research efforts for enzymatic lignin depolymerization were especially focusing on laccases (copper oxidases), as these enzymes require no cofactors, or co-substrates (such as hydrogen peroxide), they use oxygen as an electron acceptor and produce water as the only by-product. Prospective and challenges of laccase application in biotechnology were recently reviewed [28, 29]. The vast majority of industrially available laccases are fungal enzymes. These enzymes, however efficient, work in acidic-to-neutral pH [30], at which lignin is hardly soluble in water. This prevents their industrialization in this area.

Recently METGEN has developed and brought to the market a proprietary lignin refining technology METNIN™. This technology is based on combining enzyme-catalyzed lignin oxidation and cascading membrane fractionation. The enzymatic element of this technology is a proprietary artificially evolved enzyme MetZyme® METNIN™ laccase able to function under extremely alkaline conditions (typical process pH 10.5) [31, 32]. METNIN™ process is outlined in **Figure 2**. Membrane-based separation of lignin by molecular size provides useful fractions of various molecular weights.

Lignin preparations of different molecular weights can be further valorized and utilized in various industrial applications [20, 25] as long as chemical/physical properties are matching the requirements [33]. Thus, the target of lignin refining is to create lignin fractions that are bioequivalent, for example, to oil-based compounds used as resins, adhesives, composites and foams (**Table 3**).

Importantly, the absence of organic solvents in the reaction mixture allows for utilization of polymer-based ultrafiltration membranes widely used in the food industry, making this technology scalable and economically feasible. Ultrafiltration membranes of different cut-off are available and widely used in industry. The choice of membranes can be customized and adds flexibility to the technology. By adjusting process parameters, outcoming lignin properties and mass distribution between the fractions can be changed.

Demethylation is a desirable process in lignin upgrade, as it increases the number of hydroxyls and thus results in activation of lignin. Demethylation can be monitored by measuring MeOH in the reaction mixture after depolymerization using Purpald-method [34]. This is a fast and convenient method to monitor the oxidation process, however, it does not give the full picture of the chemical modification of the lignin, as some of the resulting hydroxyls can end up in new ester bonds or be further oxidized. For further characterization, titration methods and NMR need to be used. Using these methods, we observed an increased number of hydroxyls and sometimes carboxylic groups per gram of dry matter, especially towards lower Mw fractions.

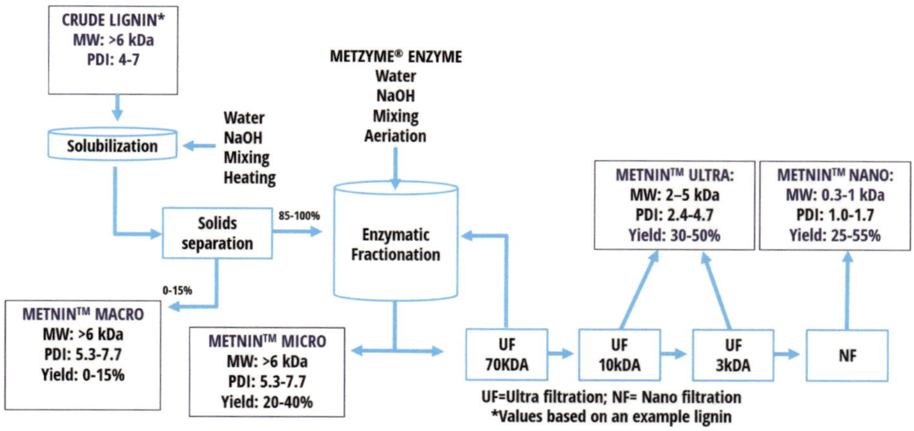

Figure 2.
METNIN™ process, schematic representation.

Lignin type	Application areas examples	Bio-equivalent of	Indicative price of the oil-based chemical
Crude Lignin	Fuel	Oil/Electricity	50–100 €/ton
METNIN™ MACRO	Bitumen, Fillers (Market established products, agro)	SBS Polymer, Inorganic fillers	400–600 €/ton
METNIN™ MICRO	Coatings & Surface treatment (Sizing value chain, Carbon fibers, agro)	Phenol resins, AKD, ASA, Wax, Latex	1000–2000 €/ton
METNIN™ ULTRA	Composites (Toy value chain)	Polyols	1500–3000 €/ton
METNIN™ NANO	Carbon Fibers (re-polymerized version), new materials	Specialty Chemicals, aromatics & phenols	> 4000 €/ton

Table 3.
METNIN™ products.

METNIN™ process allows tuning the resulting fractions in several ways: by choosing membranes with different cut-offs according to the desired molecular weight, by adjusting the extent of oxidation to tune other properties such as the content of OH-groups (phenolic aliphatic and carboxylic), and controlled polymerization, which affects reactivity and solubility of the resulting fractions. Thus variance in the starting material can be compensated by the process adjustment and the refining process results in more homogeneous fractions with a less batch-to-batch variation. Post-fractionation processing of the fractions can further tune the properties – purity and solubility in water or solvents.

Refining of lignin in METNIN™ process is accompanied by chemical activation via demethylation and benzylic oxidation as well as increased solubility in neutral and acidic pH and altered colloidal behavior. The process parameters largely depend on the starting lignin itself. In practice, each new lignin needs to be investigated to understand its behavior in the process.

One of the biggest challenges of lignin valorization is that lignin's structure is highly dependent not only on the species of wood but also on the treatment and extraction method. Therefore, the process parameters of lignin oxidation and fractionation need to be optimized experimentally. MetGen not only provides the licensing of the technology but also offers customer-specific projects for demonstration of the impact of METNIN™ process on customer's lignin.

METNIN™ process has been demonstrated and is routinely run at a pilot-scale (400 Liters reactor vessel). In addition, technology transfer to a ton (1000 kg) scale batch production has been completed and an engineering package for the industrial scale is developed and available for licensing from MetGen. Lignin fractions were tested in various applications.

Oligomeric fraction METNIN™ ULTRA was used as lignopolyol to completely replace a commercial polyol in polyurethane rigid foam formulations [35]. The specifications of the obtained foams such as closed cell count, water uptake, and compression characteristics, were all within industry standards for rigid foam applications.

METNIN™ MICRO showed excellent potential in paper coating application and the respective product is being developed together with the pulp and paper industry. Other applications are being tested with industry partners.

4.3.2 Cellulose fibers modification

Cellulose is the most traditional product from non-food biomass. Cellulose fibers further turned into paper were produced for centuries by the pulp and paper industry. However, if the printing and writing paper used to be the main product of this industry, the recent changes in the consumer market and the digitalization of the information market shifted the focus of the pulp and paper industry to hygiene and packaging products, which are much more diverse in terms of the required properties of the fibers (strength, softness, odor, water absorption/resistance etc). Changing the fiber properties can be achieved by adjusting the wood refining and chemical treatment, however, it can also be enhanced by enzymatic treatment [36]. For this purpose, individual enzymes or a set of enzymes are needed rather than a natural hydrolytic cocktail.

The main component of fiber strength improvement cocktail is endoglucanase, an enzyme that introduces individual brakes in a cellulose strand [37]. It attacks amorphous regions of cellulose fiber, where the crystalline structure was distorted by refining. These brakes make fibers more "hairy" and improve fibrillation (incorporation of the fibers into paper webbing), which eventually translates into improved strength properties of the paper [38, 39]. This enzymatic activity can also be used to even further cleave the amorphous region and help to create nano-cellulose [40], which is widely used in various applications from packaging to electronics and health. The conventional mechanical process of obtaining nanocellulose is highly energy demanding and enzymes can considerably reduce the required refining energy. Another cellulose base product with growing demand is the so-called dissolving pulp, which is used for viscose production. The process of liquefying the pulp by separating the fibrils (the strands of cellulose) is also highly energy demanding and chemically polluting. Viscose production can be more eco-friendly and economic by using enzymes, specifically xylanases and cellulases to selectively remove hemicelluloses and improve pulp reactivity, respectively [41].

These cellulose products are usually not considered to be in the scope of biorefineries but rather pulp and paper industry, however, some fiber-based products could be introduced into the biorefineries offering. The MetZyme®BRILA™ product family of MetGen's portfolio is dedicated to fiber modification solutions (see **Table 2**).

4.3.3 Glucose conversion

The main outcome of cellulose processing in the biorefineries is currently glucose. Glucose is the central nutrient in the microbial world. Almost all microorganisms can be cultivated on glucose with some supplement of nitrogen-containing compounds and microelements. Thus the demand for glucose will grow as the bioeconomy develops. And glucose can be a starting material for bioconversion to practically any natural compound by a whole-cell microbial factory.

Apart from being a central nutrient for microbial production glucose can also serve as a starting material for platform chemicals [42]. By acid catalysis, sugar molecules can be converted to platform chemicals such as hydroxy-methyl furfural (HMF), furfuryl alcohol (FAL), and levulinic acid (LA) which can be further used for polymer synthesis [43].

HMF is an important emerging platform chemical that can be further converted to 2,5-furan dicarboxylic acid (FDCA) by chemical [44] or enzymatic [45] oxidation. In turn, FDCA is a precursor for a new to the world polymer polyethylene 2,5-furandicarboxylate (PEF) which provides an alternative to the oil-based plastics polyethylene terephthalates (PET) used for the majority of disposable plastic bottles. Remarkably, PEF represents not only a biobased alternative to PET but also provides a technical advantage in gas retention, which is extremely important for carbonated drinks' shelf life. Apart from PEF, other polyesters and various polyamides and polyurethanes containing FDCA have been described in the literature [46].

HMF can be obtained by dehydration of carbohydrates. The preferred substrate for dehydration is fructose, which can be obtained by the chemical or enzymatic isomerization of glucose (**Figure 3**).

The respective enzyme is called glucose isomerase, although biochemically speaking it is xylose isomerase with a side activity of glucose isomerization [47]. These enzymes are widely available: they are one of the largest in volume in the industrial enzyme market for their production of widely-used High Fructose Syrups (HFS) for food applications. Isomerization is a reversible reaction; enzymes bring the mixture of glucose and fructose to the equilibrium ratio of about 1:1 and the reaction stops. Fructose provides sweetness for food and beverages, but can also be used as an intermediate for bioplastics.

The currently available commercial enzymes are highly sensitive to substrate sugar impurities. This is acceptable for food industry applications, where sugar has to be pure anyway. Typically, even sugar produced from starch requires activated carbon filtration, ion-exchange chromatography, and degasification before it can proceed to isomerization reaction as described by the technology providers, see, for example https://www.myandegroup.com/starch-syrup-process-technology.html. Sugars produced from 2nd generation biorefinery (especially from wood) have much more impurities than starch-derived sugar, including lignin, extractives, etc. It would need a lot of purification to enable utilization of currently available glucose isomerase, and the required level of purity is not justified for the technical sugar. MetGen has developed a proprietary industrially relevant recombinant bacterial glucose isomerase with high tolerance to substrate impurities. The enzyme can work directly in biomass hydrolysate. Lower requirements for purification lead to reduced process costs. This enzyme was further engineered to be much less sensitive to the presence of xylose - a preferred substrate of all of the natural glucose isomerases, and thus a potent competitive inhibitor in the reaction with glucose.

Proposed enzymatic solutions for biorefinery and especially for sugars-to-biochemicals pathways are numerous, however, they are still mostly in the research and development stage [48], and the takeoff of the bio-based economy

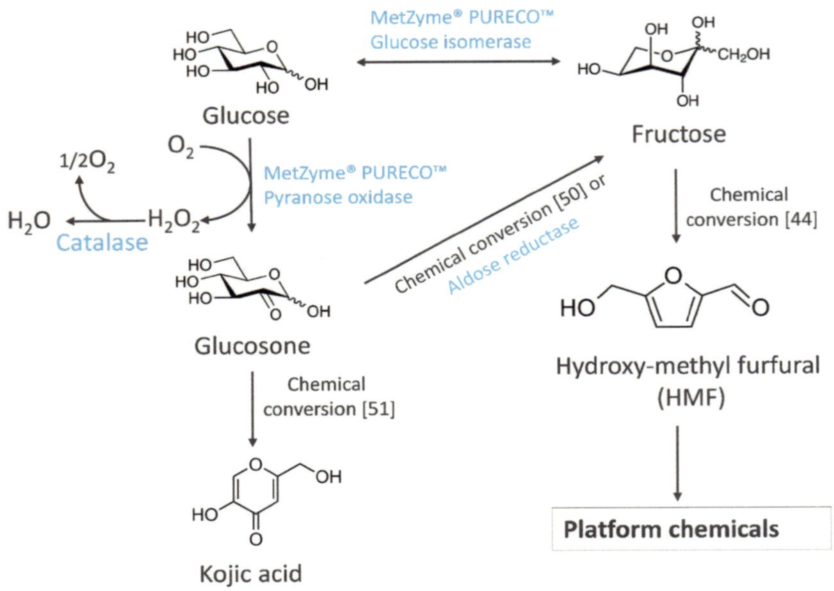

Figure 3.
Bioconversion of glucose using MetZymes® and related processes.

largely depends on the success of this effort. MetZyme® PURECO™ product family – to which previously mentioned glucose isomerase also belongs – is dedicated to next-generation enzymes allowing specific conversions towards high value-addition renewable chemicals, which are beyond sugars. One of these enzymes MetZyme® PURECO™Pyranose oxidase opens an economical route to previously commercially unavailable above a gram scale compound glucosone (2-Keto-D-glucose). While until now the use of d-glucosone has been limited due to its high price and limited availability, it has been envisaged for a long time to provide an alternative route to fructose [49]. As opposed to isomerization reaction, oxidation of glucose to glucosone can be driven to completion, and glucoson's aldehyde group can be further reduced with high specificity to a hydroxyl leading to fructose (**Figure 3**). The second step can be performed either by chemical hydrogenation [50] or enzymatically by an aldose reductase [49]. Recently, more applications of glucosone started to be developed, for example, it has been shown, that certain fine chemicals such as kojic acid could potentially be produced from this source (**Figure 3**) [51].

MetZyme®PURECO™ Pyranose oxidase and glucose isomerase are commercially available proprietary enzymes developed by MetGen in the course of the Horizon 2020 research and innovation program, funded by the European Union's Bio-Based Industries Joint Undertaking.

5. Concluding remarks

MetGens philosophy in serving biobased industries is to provide a full solution rather than on-shelf enzymes to the customers, and where possible an engineering package. Therefore, MetGen embraces all stages of enzyme technology development from enzyme discovery and molecular biology to application testing, streamlined efficient production, and integration into an industrial process. We call this ENZYNE platform.

Another vastly important principle for us is to take an active part in open innovation to combine forces with industry players to build the new value chains in the bioeconomy.

Building consumer awareness is also key for the expansion of the bioeconomy. Society needs to gain a common understanding of the importance of bio-based solutions and their impact on sustainability and circularity.

Acknowledgements

This research recieved funding from the Bio Based Industries Joint Undertaking under the European Union's Horizon 2020 research and innovation program under grant agreement No: 792061 SWEETWOODS is acknowledged.

Author details

Klara Birikh[1*], Alex Michine[2], Matti Heikkilä[2] and Petri Ihalainen[2]

1 MetGen Oy, Geleen, The Netherlands

2 MetGen Oy, Kaarina, Finland

*Address all correspondence to: klara@metgen.com

References

[1] Huang K, Fasahati P, Maravelias CT. System-Level Analysis of Lignin Valorization in Lignocellulosic Biorefineries. iScience. 2020;23(1):100751. Available from: https://doi.org/10.1016/j.isci.2019.100751

[2] Escamilla-Alvarado C, Pérez-Pimienta JA, Ponce-Noyola T, Poggi-Varaldo HM. An overview of the enzyme potential in bioenergy-producing biorefineries. Vol. 92, Journal of Chemical Technology and Biotechnology; 2017. p. 906-924. Available from: https://onlinelibrary.wiley.com/doi/full/10.1002/jctb.5088

[3] Woodyer RD, Johannes TW, Zhao H. Regeneration of Cofactors for Enzyme Biocatalysis. Enzyme Technology. Springer. New York; 2006. p. 85-103. Available from: https://link.springer.com/chapter/10.1007/978-0-387-35141-4_5

[4] Liu W, Wang P. Cofactor regeneration for sustainable enzymatic biosynthesis. Vol. 25, Biotechnology Advances Biotechnol Adv; 2007. p. 369-384. Available from: https://pubmed.ncbi.nlm.nih.gov/17459647/

[5] Shakeel T, Sharma A, Yazdani SS. Building cell factories for the production of advanced fuels. Vol. 47, Biochemical Society Transactions. 2019. p. 1701-1714. Available from: https://pubmed.ncbi.nlm.nih.gov/31803925/

[6] Lutz S, Iamurri SM. Protein engineering: Past, present, and future. Methods in Molecular Biology. Humana Press Inc.; 2018. p. 1-12. Available from: https://pubmed.ncbi.nlm.nih.gov/29086300/

[7] Ko YS, Kim JW, Lee JA, Han T, Kim GB, Park JE, et al. Tools and strategies of systems metabolic engineering for the development of microbial cell factories for chemical production. Vol. 49, Chemical Society Reviews. Royal Society of Chemistry; 2020. p. 4615-4636. Available from: https://pubmed.ncbi.nlm.nih.gov/32567619/

[8] Galbe M, Wallberg O. Pretreatment for biorefineries: A review of common methods for efficient utilization of lignocellulosic materials. Vol. 12, Biotechnology for Biofuels; 2019. p. 1-26. https://doi.org/10.1186/s13068-019-1634-1

[9] Baruah J, Nath BK, Sharma R, Kumar S, Deka RC, Baruah DC, et al. Recent trends in the pretreatment of lignocellulosic biomass for value-added products. Vol. 6, Frontiers in Energy Research. 2018. p. 141. Available from: www.frontiersin.org

[10] Bischof RH, Ramoni J, Seiboth B. Cellulases and beyond: the first 70 years of the enzyme producer Trichoderma reesei. Microb Cell Fact. 2016 Jun 10;15(1):106. Available from: http://microbialcellfactories.biomedcentral.com/articles/10.1186/s12934-016-0507-6

[11] Shida Y, Furukawa T, Ogasawara W. Deciphering the molecular mechanisms behind cellulase production in *Trichoderma reesei*, the hyper-cellulolytic filamentous fungus. Biosci Biotechnol Biochem. 2016 Sep 14;80(9):1712-1729. Available from: http://www.tandfonline.com/doi/full/10.1080/09168451.2016.1171701

[12] Hsieh CWC, Cannella D, Jørgensen H, Felby C, Thygesen LG. Cellulase inhibition by high concentrations of monosaccharides. J Agric Food Chem. 2014 Apr 30;62(17):3800-3805. Available from: https://pubs.acs.org/doi/abs/10.1021/jf5012962

[13] Yang Y, Zhang X, Yin Q, Fang W, Fang Z, Wang X, et al. A mechanism of glucose tolerance and stimulation of GH1 β-glucosidases. Sci Rep. 2015;5(November):1-12.

[14] Romero-Téllez S, Lluch JM, González-Lafont À, Masgrau L. Comparing hydrolysis and transglycosylation reactions catalyzed by Thermus thermophilus β-glycosidase. A combined MD and QM/MM study. Front Chem. 2019 Apr 10;7(APR):200. Available from: www.frontiersin.org

[15] Xu J, Zhao G, Kou Y, Zhang W, Zhou Q, Chen G, et al. Intracellular β-glucosidases CEL1a and CEL1b are essential for cellulase induction on lactose in Trichoderma reesei. Eukaryot Cell. 2014 Aug 1;13(8):1001-1013. Available from: https://pubmed.ncbi.nlm.nih.gov/24879125/

[16] Zhang J, Qu Y, Xiao P, Wang X, Wang T, He F. Improved biomass saccharification by Trichoderma reesei through heterologous expression of lacA gene from Trametes sp. AH28-2. J Biosci Bioeng. 2012 Jun 1;113(6):697-703. Available from: https://www.sciencedirect.com/science/article/pii/S1389172312000576

[17] Selig MJ, Knoshaug EP, Decker SR, Baker JO, Himmel ME, Adney WS. Heterologous expression of Aspergillus niger β-D-xylosidase (XlnD): Characterization on lignocellulosic substrates. Appl Biochem Biotechnol. 2008 Mar;146(1-3):57-68. Available from: https://pubmed.ncbi.nlm.nih.gov/18421587/

[18] Selig MJ, Knoshaug EP, Adney WS, Himmel ME, Decker SR. Synergistic enhancement of cellobiohydrolase performance on pretreated corn stover by addition of xylanase and esterase activities. Bioresour Technol. 2008;99(11):4997-5005. Available from: https://pubmed.ncbi.nlm.nih.gov/18006303/

[19] Zhang F, Bunterngsook B, Li J-X, Zhao X-Q, Champreda V, Liu C-G, et al. Regulation and production of lignocellulolytic enzymes from Trichoderma reesei for biofuels production. Advances in Bioenergy; 2019. p. 79-119.

[20] Laurichesse S, Avérous L. Chemical modification of lignins: Towards biobased polymers. Prog Polym Sci. 2014;39(7):1266-1290. Available from: http://dx.doi.org/10.1016/j.progpolymsci.2013.11.004

[21] Schutyser W, Renders T, Van Den Bosch S, Koelewijn SF, Beckham GT, Sels BF. Chemicals from lignin: An interplay of lignocellulose fractionation, depolymerisation, and upgrading. Chem Soc Rev. 2018;47(3):852-908. Available from: http://dx.doi.org/10.1039/C7CS00566K

[22] Sun Z, Fridrich B, De Santi A, Elangovan S, Barta K. Bright Side of Lignin Depolymerization: Towards New Platform Chemicals. Vol. 118, Chemical Reviews. 2018. p. 614-678. Available from: https://pubs.acs.org/sharingguidelines

[23] Rinaldi R, Jastrzebski R, Clough MT, Ralph J, Kennema M, Bruijnincx PCA, et al. Paving the Way for Lignin Valorisation: Recent Advances in Bioengineering, Biorefining and Catalysis. Angew Chemie - Int Ed. 2016;55(29):8164-8215.

[24] Kim YS, Kadla JF. Preparation of a thermoresponsive lignin-based biomaterial through atom transfer radical polymerization. Biomacromolecules. 2010;11(4):981-988.

[25] Gandini A, Belgacem MN. Lignins as components of macromolecular materials. In: Gandini A, Belgacem MN, editors. Monomers, polymers and composites from renewable resources. Elsevier Science; 2008. p. 243-271.

[26] Xing Q, Ruch D, Dubois P, Wu L, Wang W-J. Biodegradable and High-Performance Poly(butylene adipate- *co* -terephthalate)–Lignin UV-Blocking Films. ACS Sustain Chem Eng. 2017;acssuschemeng.7b02370. Available from: http://pubs.acs.org/doi/abs/10.1021/acssuschemeng.7b02370

[27] Wong DWS. Structure and Action Mechanism of Ligninolytic Enzymes. Vol. 157, Applied Biochemistry and Biotechnology. 2009. 174-209 p. Available from: http://link.springer.com/10.1007/s12010-008-8279-z

[28] Mate DM, Alcalde M. Laccase: A multi-purpose biocatalyst at the forefront of biotechnology. Microb Biotechnol. 2017;10(6): 1457-1467.

[29] Roth S, Spiess AC. Laccases for biorefinery applications: a critical review on challenges and perspectives. Bioprocess Biosyst Eng. 2015 Dec;38(12):2285-2313.

[30] Rodgers CJ, Blanford CF, Giddens SR, Skamnioti P, Armstrong FA, Gurr SJ. Designer laccases: a vogue for high-potential fungal enzymes? Trends Biotechnol. 2010;28(2):63-72.

[31] Hämäläinen V, Grönroos T, Suonpää A, Heikkilä MW, Romein B, Ihalainen P, et al. Enzymatic Processes to Unlock the Lignin Value. Front Bioeng Biotechnol. 2018;6(March):1-10. Available from: http://journal.frontiersin.org/article/10.3389/fbioe.2018.00020/full

[32] Wang L, Tan L, Hu L, Wang X, Koppolu R, Tirri T, et al. On Laccase-Catalyzed Polymerization of Biorefinery Lignin Fractions and Alignment of Lignin Nanoparticles on the Nanocellulose Surface via One-Pot Water-Phase Synthesis. ACS Sustain Chem Eng. 2021;acssuschemeng.1c01576 Available from: https://pubs.acs.org/doi/10.1021/acssuschemeng.1c01576

[33] Passoni V, Scarica C, Levi M, Turri S, Griffini G. Fractionation of Industrial Softwood Kraft Lignin: Solvent Selection as a Tool for Tailored Material Properties. ACS Sustain Chem Eng. 2016;4(4):2232-2242.

[34] Anthon GE, Barrett DM. Comparison of Three Colorimetric Reagents in the Determination of Methanol with Alcohol Oxidase. Application to the Assay of Pectin Methylesterase. J Agric Food Chem. 2004 Jun;52(12):3749-3753.

[35] Antunes J, Levée T, Ihalainen P. Application of Enzymatically Treated Lignin Oligomers as Lignopolyols For A Full Replacement of Commercial Polyols in Polyurethane Foam Formulation. Biomed J Sci Tech Res. 2019;24(1).

[36] Birikh KR, Heikkila MW, Michine A, Mialon A, Grönroos T, Ihalainen P, et al. Metgen: Value from Wood - Enzymatic Solutions. In: de Gonsalo G, Dominguez de Maria P, editors. Biocatalysis: An Industrial Perspective (Catalysis Series). The Royal Society of Chemistry; 2017. p. 298-322.

[37] Yang B, Dai Z, Ding S-Y, Wyman CE. Enzymatic hydrolysis of cellulosic biomass. Biofuels. 2011 Jul 9;2(4):421-449. Available from: https://www.tandfonline.com/doi/abs/10.4155/bfs.11.116

[38] Wang M, Du J, Zhang D, Li X, Zhao J. Modification of Different Pulps by Homologous Overexpression Alkali-Tolerant Endoglucanase in Bacillus subtilis Y106. Sci Rep. 2017 Dec 12;7(1):3321. Available from: http://www.nature.com/articles/s41598-017-03215-9

[39] Han X, Bi R, Khatri V, Oguzlu H, Takada M, Jiang J, et al. Use of Endoglucanase and Accessory Enzymes to Facilitate Mechanical Pulp Nanofibrillation. ACS Sustain Chem Eng. 2021 Jan 25;9(3):1406-1413. Available from: https://pubs.acs.org/doi/abs/10.1021/acssuschemeng.0c08588

[40] Alonso-Lerma B, Barandiaran L, Ugarte L, Larraza I, Reifs A, Olmos-Juste R, et al. High performance crystalline nanocellulose using an ancestral endoglucanase. Commun Mater. 2020 Dec 12;1(1):1-10. Available from: https://doi.org/10.1038/s43246-020-00055-5

[41] Kaur P, Bhardwaj NK, Sharma J. Application of Microbial Enzymes in Dissolving Pulp Production. In: Frontier Discoveries and Innovations in Interdisciplinary Microbiology. New Delhi: Springer India; 2016. p. 133-156. Available from: https://link.springer.com/chapter/10.1007/978-81-322-2610-9_8

[42] Takkellapati S, Li T, Gonzalez MA. An overview of biorefinery-derived platform chemicals from a cellulose and hemicellulose biorefinery. Clean Technol Environ Policy. 2018 Sep 1;20(7):1615-1630

[43] Assary RS, Kim T, Low JJ, Greeley J, Curtiss LA. Glucose and fructose to platform chemicals: Understanding the thermodynamic landscapes of acid-catalyzed reactions using high-level ab initio methods. Phys Chem Chem Phys. 2012 Dec 28;14(48):16603-16611. Available from: https://pubmed.ncbi.nlm.nih.gov/22932938/

[44] Van Putten RJ, Van Der Waal JC, De Jong E, Rasrendra CB, Heeres HJ, De Vries JG. Hydroxymethylfurfural, a versatile platform chemical made from renewable resources. Chem Rev. 2013;113(3):1499-1597.

[45] Cajnko MM, Novak U, Grilc M, Likozar B. Enzymatic conversion reactions of 5-hydroxymethylfurfural (HMF) to bio-based 2,5-diformylfuran (DFF) and 2,5-furandicarboxylic acid (FDCA) with air: Mechanisms, pathways and synthesis selectivity. Biotechnol Biofuels. 2020 Apr 10 13(1):66. Available from: https://doi.org/10.1186/s13068-020-01705-z

[46] Sousa AF, Vilela C, Fonseca AC, Matos M, Freire CSR, Gruter G-JM, et al. Biobased polyesters and other polymers from 2,5-furandicarboxylic acid: a tribute to furan excellency. Polym Chem. 2015;6(33):5961-5983. Available from: http://xlink.rsc.org/?DOI=C5PY00686D

[47] Bhosale SH, Rao MB, Deshpande V V. Molecular and industrial aspects of glucose isomerase. Microbiol Rev. 1996;60(2):280-300. Available from: http://mmbr.asm.org/content/60/2/280.long

[48] Hasunuma T, Okazaki F, Okai N, Hara KY, Ishii J, Kondo A. A review of enzymes and microbes for lignocellulosic biorefinery and the possibility of their application to consolidated bioprocessing technology. Bioresour Technol. 2013 May;135:513-522. Available from: https://pubmed.ncbi.nlm.nih.gov/23195654/

[49] Leitner C, Neuhauser W, Volc J, Kulbe KD, Nidetzky B, Haltrich D. The Cetus process revisited: A novel enzymatic alternative for the production of aldose-free D-fructose. Biocatalysis and Biotransformation. Harwood Academic Publishers GmbH; 1998. p. 365-382. Available from: https://www.tandfonline.com/doi/abs/10.3109/10242429809003629

[50] Lassfolk R, Aho A, Murzin DY, Leino R. Hydrogenation of crude and purified d-glucosone generated by enzymatic oxidation of d-glucose.

RSC Adv. 2020 Aug 18;10(51):30476-30480. Available from: https://pubs.rsc.org/en/content/articlehtml/2020/ra/d0ra05512c

[51] Lassfolk R, Suonpaä A, Birikh K, Leino R. Chemo-enzymatic three-step conversion of glucose to kojic acid. Chem Commun. 2019 Dec 5;55(98):14737-14740. Available from: https://pubs.rsc.org/en/content/articlehtml/2019/cc/c9cc07405h

Chapter 4

Progress on the Co-Pyrolysis of Coal and Biomass

Xiye Chen, Linyao Zhang, Li Liu, Chang Xing, Yan Zhao, Kirk Cobb, Roger Ruan and Penghua Qiu

Abstract

In this chapter, the synergistic mechanism and the resulting influence during co-pyrolysis of coal and biomass, are summarized. The properties of coal and biomass, the release and migration of alkali and alkaline earth metals (AAEMs), the interaction between volatile and char, the characteristics of the resulting volatiles, and the physicochemical structure and reactivity of co-pyrolysis char, are also analyzed. In addition, the influence of AAEMs on the properties of the co-pyrolysis products is reviewed. Moreover, the analysis of the co-pyrolysis industry demonstration is also mentioned. Finally, this chapter also proposes some additional possibilities, based on further literature research.

Keywords: Co-pyrolysis, Coal/Biomass, Characteristics of co-pyrolysis products, Volatiles-char interaction, Alkali and alkaline earth metals

1. Introduction

Energy supply is the fundamental basis for rapid economic growth and sustainable social development. Due to abundant reserves of coal worldwide, it has been become one of the most important fossil fuels of the past two centuries, and may continue to be used, somewhat, for up to an additional 200years in the future [1]. However, the use of coal can cause serious environmental problems. For example, sulfur dioxide and nitrogen oxides produced while burning coal can pollute the air and water, and are extremely harmful to humans, animals, and other organisms. In addition, coal burning is one of the main sources of greenhouse gas (GHG) emissions, accounting for at least 20–30% of total CO_2 emissions [2].

Biomass energy, resulting from natural photosynthesis, represents a renewable form of energy, and is the fourth largest energy source after coal, oil, and nature gas [3]. It can supply about 14% of the world's energy consumption, and about 38% of the energy consumption in developing countries [4, 5]. Because of its green, low-carbon, clean and renewable characteristics, it has become one of the most important sustainable energy sources [6, 7]. In order to actively respond to climate change, China has announced that it intends to reach its national CO_2 emission peak before 2030, and thereafter achieve carbon neutrality before 2060 [8]. Under this background, biomass energy, as a zero-carbon energy source, will play an important role, and the biomass energy industry will usher in a period of major development opportunities. At present, China produces about 6.3 billion tons of various organic wastes (including: agricultural and forestry residues, domestic waste, domestic sludge, livestock and poultry manure, fruit and vegetable residues, and industrial organic waste liquids) every year; this organic waste is equivalent to about 800 million tons of standard coal. Of this total amount of organic waste, it is estimated that the amount of biomass resources that can be utilized as an energy source in China each year, is approximately equal to 460 million standard tons of coal [9]. Among these, the amount of agricultural waste is about 400 million tons, which is equivalent to about 200 million tons of standard coal. The amount of forestry waste is about 350 million tons, which is also equivalent to about 200 million tons of standard coal. The remaining organic wastes are equivalent to about 60 million tons of standard coal. The development of biomass energy can result in the utilization of eco-friendly, sustainable energy, and also reduce pollution due to the inappropriate discarding of organic wastes; it can also realize the utilization of an otherwise ignored energy resource. More importantly, the utilization of biomass energy cannot only assist in achieving carbon neutrality, but can also result in "negative" carbon emissions when biomass energy is utilized via carbon capture and storage (BECCS). However, biomass has its drawbacks, such as seasonal harvest rather than year-round availability, wide distribution, low energy density, and high transportation costs [10]. These shortcomings, especially the limited supply of biomass raw materials, currently restrict its large-scale industrial application in China. Unfortunately, China's current utilization of organic waste energy is less than 5%. But if China can expand its use of available biomass, it can begin to reduce its use of coal.

The co-utilization of coal and biomass can not only reduce the pressure of coal supply and environmental problems, but also save the cost of building direct biomass utilization equipment. In terms of fuel characteristics, coal and biomass also have a great possibility to complement one another. In order to utilize them on a large, efficient scale, the co-utilization of coal and biomass may offer a potential benefit, as a promising technical method. As the initial stage of thermal chemical conversion, the co-pyrolysis process of coal and biomass is very important, since it determines the formation characteristics, structures, and properties of volatiles (gas products), tar (liquid products) and char (solid products). The main components of biomass pyrolysis are volatiles and tar; in comparison, the main component of coal pyrolysis is char, which can reach 40–60%. When biomass is combined with coal, the yield of char is affected by the ratio during the co-pyrolysis of biomass and coal. The reaction of the solid phase product (co-pyrolysis char) is the slowest step in the whole thermochemical conversion process, and its reaction rate determines the rate of the whole thermochemical reaction. On the one hand, during co-pyrolysis, the volatiles produced from the biomass and coal, can interact with the co-pyrolysis char, which leads to changes in the properties of the resulting char and volatiles. On the other hand, the changes in the properties of the char and volatiles, can also affect the interactions between volatiles and co-pyrolysis char. The two are interrelated and influence each other. Therefore, during co-pyrolysis, it is very important to study the interactions between the resulting volatiles and char.

Compared to coal, biomass contains more CaO, K_2O, P_2O_5, MgO, Na_2O and Mn, and less SiO_2, Al_2O_3, Fe_2O_3, SO_3 and TiO_2. Among these, K^+, Na^+, Ca^{2+} and Mg^{2+} belong to alkali and alkaline earth metals (AAEMs) [11]. The content of the alkali and alkaline earth metal oxides in biomass ash exceeds 27%, while the content in coal ash is only 6–10%. The content of silicon and aluminum oxides in biomass ash is 22–57%, while the content in coal ash exceeds 80% [12, 13]. AAEMs play an important role in the process of coupling utilization of coal and biomass, and are good catalysts for combustion and gasification reactions, which can significantly affect the reactivity of the resulting co-pyrolysis char [14]. The presence of AAEMs can affect the dynamic pyrolysis process, and has a direct catalytic effect on the cracking of volatiles and their precursors. However, due to the diversity and superimposition of the reactions, the nature of the interactions between coal and biomass during co-pyrolysis has not been fully understood, especially the catalytic influence of AAEMs, including the influence mechanism of AAEMs on the volatiles generation, the influence of AAEMs on the interaction between volatiles and char, and the influence of AAEMs on the co-pyrolysis char reaction. Therefore, the chemical mechanism of AAEMs during co-pyrolysis is one of the key issues that needs to be investigated further, regarding the basic research of coal and biomass co-utilization.

In this chapter, a comprehensive overview of the co-pyrolysis of coal and biomass is presented. The focus of interest is mainly on the chemical mechanisms, during co-pyrolysis. The properties of coal and biomass, the synergistic mechanism, the release and migration of AAEMs, the interactions between volatiles and char, the volatiles production characteristics, the physicochemical structures and reactivity of co-pyrolysis char, are analyzed in this chapter. Moreover, the influence of AAEMs on the properties of the co-pyrolysis products, is also presented.

2. Properties of coal and biomass

Coal is an extremely complex and heterogeneous mixture composed of organic macromolecules and inorganic minerals [11]. It was formed by ancient plants, buried in the ground and experienced complex chemical changes at high temperatures and pressures. The transformation process involved the loss of hydrogen and oxygen and the condensation of carbon. "Coal" can be divided into peat, lignite, bituminous, and anthracite coals, according to the stage of formation and degree of coalification. Although peat is fuel, it is not actually coal, but a "pre-coal". The main components of the organic macromolecular networks of coal are carbon, hydrogen, oxygen, nitrogen, and sulfur. Calculated by weight, carbon is the main component, accounting for 60% - 95% of the total weight. The carbon content of most coals is below 90%, and the hydrogen content is generally around 5%, while the hydrogen content of coals containing 95% carbon drops to about 2% [15]. The nitrogen content of coals is generally between 1% - 2%. Sulfur is also a very important component of most coals, accounting for 1–4%. The oxygen content is inversely proportional to the carbon content, that is, the higher the carbon content in coal, the lower the oxygen content. The oxygen content in coal is important because coal with more oxygen is more likely to catch fire. Carbon in coal mainly exists in two forms, namely fixed carbon or volatile matter. The ratio of fixed carbon to volatile matter determines the rank of coal [16]. Inorganic minerals account for a small proportion in the overall composition of coal. However, AAEMs in the minerals have obvious catalytic effects on the thermochemical conversion reaction of coal [14, 17].

A wide range of biomass fuels are available in the environment, ranging from wood to materials derived from herbaceous plants and straw. Usually, biomass fuels are classified according to their source and properties. Biomass can be divided into primary residues, secondary residues, tertiary residues, and energy crops according to their sources [18]. Primary residues include biomass such as wood, straw, grain, and corn, which are usually obtained as by-products from forest products and food crops [19]. Secondary residues are derived from biomass materials used in industrial products and food production, such as sawmills, paper mills, food and beverage industries, apricot and other fruit seeds. Tertiary residues include waste materials and dismantled timber, from other previously used biomass materials [20]. Energy crops may include willow, poplar, switchgrass, and miscanthus grass. In addition, biomass can be divided into four types based on properties: woody biomass, herbaceous biomass, organic waste, and aquatic biomass (such as kelp) [21]. Among all these types

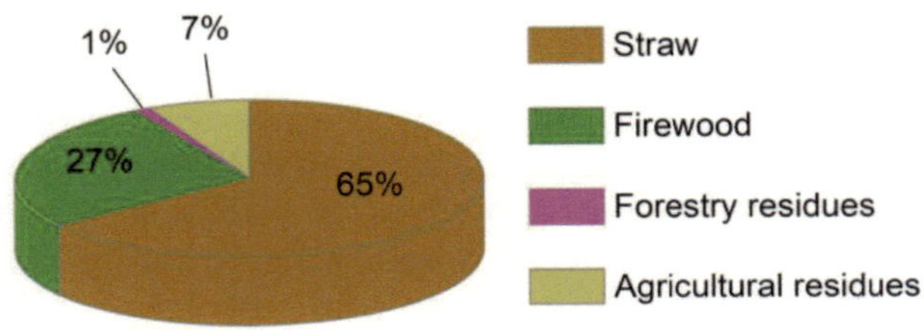

Figure 1.
The proportions of four types biomass with the largest reserves.

of biomass, the four types of biomass with the largest reserves are straw, firewood, forestry residues, and agricultural residues; their proportions are shown in **Figure 1**.

Biomass is mainly composed of cellulose, hemicellulose, and lignin. Generally, lignocellulosic biomass contains about 35–55% cellulose, 20–35% hemicellulose, and 10–30% lignin [15]. Cellulose is a linear polymer formed by the connection of glucose molecules through ether bonds, which is the most abundant carbohydrate in nature and the main component of plant plasm cells. It decomposes in the temperature range of 240–350°C. Hemicellulose is a mixture of different polysaccharides with a low degree of polymerization and no crystal structure, so it is easily hydrolyzed. The thermal degradation of hemicellulose occurs at temperatures between 130°C and 260°C, mainly above 180°C. Lignin is composed of hydroxyphenyl propane, guaiacyl propane and syringyl propane. These monomers are formed by disordered combination of C-C bonds and C-O bonds through dehydrogenation polymerization. Lignin decomposes over a wide temperature range of 280–500°C.

There are significant differences between biomass and coal in proximate analysis, ultimate analysis, calorific value, ash composition, physical structure, chemical structure, and reactivity. The key differences in the properties of biomass in comparison with coal (see **Tables 1** and **2**) are [6, 11, 24–26]: (1) more moisture and volatiles, less fixed carbon and ash content; (2) more O, H and Cl, less C, N and S; (3)

Fuel	Ultimate analysis (db.% w/w)					Proximate analysis (% w/w)			LHV (MJ/kg)	
	C	H	O	N	S	Ash	VM	FC	M	
Rice husk	35.20	4.79	59.00	1.01	—	9.40	66.12	13.80	10.65	13.12
Bamboo dust	43.45	5.49	50.74	0.33	—	2.68	70.83	15.62	10.87	14.85
Wood Sawdust	42.30	5.17	51.73	0.80	—	1.40	69.29	17.84	11.37	12.86
Cedar wood	51.10	5.90	42.50	0.12	0.02	0.30	80–82	18–20	—	19.26
Olive–oil residue	50.70	5.89	36.97	1.36	0.30	4.60	76.00	19.40	9.50	21.20
Rice straw	38.61	4.28	37.16	1.08	0.65	12.64	65.26	16.55	5.58	14.40
Pine sawdust	50.54	7.08	41.11	0.15	0.57	0.55	82.29	17.16	—	20.54
Spruce wood pellet	49.30	5.90	44.40	0.10	—	0.30	74.20	17.10	8.40	18.50
Marc of grape	49.66	5.56	34.42	2.23	0.14	7.83	65,77	26.40	—	19.51
Coffee husk	46.80	4.90	47.10	0.60	0.60	1.00	74.30	14.30	10.40	16.54
Coffee ground	52.97	6.51	36.62	2.80	0.05	1.00	71.80	16.70	10.50	22.00
Larch wood	44.18	6.38	49.32	0.12	—	0.12	76.86	14.86	8.16	19.45
Grapevine Pruning waste	46.97	5.80	44.49	0.67	0.01	2.06	78.16	19.78	—	17.91

Fuel	Ultimate analysis (db.% w/w)					Proximate analysis (% w/w)			LHV (MJ/kg)	
	C	H	O	N	S	Ash	VM	FC	M	
Jute stick	49.79	6.02	41.37	0.19	0.05	0.62	76–78	21–23	—	19.66
Sugar–cane bagasse	48.58	5.97	38.94	0.20	0.05	1.26	67–70	29–31	—	19.05
Corn cob	40.22	4.11	42.56	0.39	0.04	2.97	71.21	16.11	9.71	16.65
Peach stone	51.97	5.76	40.70	0.79	0.01	0.65	81.30	18.10	8.53	21.60
Wheat straw	46.10	5.60	41.70	0.50	0.08	6.01	75.80	18.10	—	17.20
Cotton stem	42.80	5.30	38.50	1.00	0.20	4.30	72.30	15.50	7.90	15.20
Straw	36.55	4.91	40.70	0.55	0.14	8.61	64.98	17.91	8.50	14.60
Camphor wood	43.43	4.84	38.53	0.32	0.10	0.49	72.47	14.75	12.29	17.48
Beech wood	48.27	6.36	45.20	0.14	—	0.80	81.00	18.00	—	19.20
Switchgrass	47.00	5.30	41.40	0.50	0.10	4.60	58.40	17.10	20.00	18.70
Petroleum coke	92.30	3.40	0.70	0.95	1.17	1.40	6.00	92.10	0.50	36.20
Lignite coal	44.66	3,66	13.90	1.0	0.21	18.42	35.17	28.27	18.4	18.05
Bituminous coal	74.73	4.43	13.68	1.02	0.19	4.08	36.95	56.90	2.07	28.05
Lean coal	66.05	3.25	2.53	1.17	0.19	25.30	20.65	53.15	0.92	24.14
Quinsam mine coal	80.30	5.50	12.60	0.9	0.70	12.90	38.80	49.10	4.20	26.99
Sub-bituminous coal	73.10	4.30	21.10	1.0	0.40	30.50	31.30	38.30	17.5	20.10
Indonesian coal	72.13	6.67	19.58	1.40	0.22	8.39	36.84	42.36	12.42	20.79
Anthracite coal	86.56	4.90	6.20	1.70	0.61	13.71	31.71	54.58	0.34	26.00
Shenmu coal	70.35	4.56	10.53	1.04	0.55	9.19	28.51	58.52	3.78	27.08
Assam coal	61.37	5.27	28.18	0.94	4.24	10.0	40.50	47.50	2.00	22.55

db: dried basis.

Table 1.
Ultimate and proximate analysis of different coal and biomass [22].

Fuel	Ash composition (wt. %)									
	SiO_2	Al_2O_3	CaO	Fe_2O_3	K_2O	MgO	Na_2O	P_2O_5	SO_3	TiO_2
Switchgrass (Manitoba, Canada)	52.50	2.10	6.40	0.30	20.30	6.50	1.60	5.00	0.02	2.60
Rice straw (Hubei Province, China)	51.99	0.91	7.68	0.84	17.61	2.33	0.96	2.49	0.04	6.50
Sawdust (Hubei Province, China)	16.47	6.50	24.89	4.57	7.76	5.56	12.84	2.42	0.58	7.64
Pine biomass (Statoil, Norway)	12.80	1.00	33.00	1.70	23.20	5.40	1.70	5.30	—	—
Corn stalks (Heilongjiang Province, China)	29.03	0.83	14.34	1.26	29.41	18.38	0.60	3.00	1.60	0.02
Sub-bituminous coal (Genesee, Alberta, Canada)	57.60	23.60	5.60	2.80	0.80	1.30	2.60	0.10	0.50	2.30

Fuel	Ash composition (wt. %)									
	SiO_2	Al_2O_3	CaO	Fe_2O_3	K_2O	MgO	Na_2O	P_2O_5	SO_3	TiO_2
Lignite coal (Inner Mongolia, China)	65.79	14.73	4.33	2.67	1.71	1.44	1.04	0.97	0.50	6.67
Bituminous coal (NSW, Australia)	47.90	26.50	7.90	7.50	0.20	0.60	0.10	1.30	1.90	6.10
Lean coal (Inner Mongolia, China)	53.99	28.44	4.07	3.22	1.56	0.88	2.97	0.97	1.82	4.00
Sub-bituminous coal (Shaanxi province, China)	53.85	11.55	13.94	10.96	0.79	1.38	0.13	0.47	2.65	0.74

Table 2.
Ash composition analysis of different coal and biomass (wt%) [22, 23].

lower calorific value; (4) higher alkali content (especially the herbaceous biomass); (5) lower bulk density, larger specific surface area, more abundant pore structure; (6) more oxygen-containing functional groups (hydroxyl, carboxyl, ether and ketone) with highly reactive groups (–COOH, $-OCH_3$ and –OH), complexes, light hydrocarbons, carbohydrates, hydroxyl oxides, carbonates, chloride and phosphate, and lower aromatics, functionality, silicate and sulfide; (7) higher reactivity.

3. Analysis of the synergistic mechanism of the co-pyrolysis of coal and biomass

The differences in the characteristics of coal and biomass determines their different pyrolysis characteristics. During the pyrolysis process, volatiles and char can interact with each other, and AAEMs can also be released and migrated. Coal and biomass are mixed during pyrolysis, so the volatiles, char and released AAEMs from the pyrolysis of both are also mixed. As a result, there may be synergies between coal and biomass during co-pyrolysis. The synergies may be caused by several factors. First, the H/C ratio of biomass is higher than that of coal, so H_2, OH and H radicals generated by biomass pyrolysis can migrate to the surface of coal during co-pyrolysis. Additional hydrogen donors may prevent the recombination and cross-linking reaction of free radicals, thus promoting coal decomposition to produce more volatiles [27]. Second, the content of AAEMs in biomass is higher than that in coal, especially the alkali metals [28]. AAEMs in biomass can migrate to the coal matrix during co-pyrolysis, and AAEMs can catalyze the pyrolysis and gas phase reaction of the coal [29]. Third, the heat transfer between coal and biomass may also cause synergistic effects during co-pyrolysis [30]. **Figure 2** shows the factors leading to synergies during co-pyrolysis of coal and biomass.

Typical pyrolysis temperatures for coal are between 350 °C and 650°C, while that of biomass are between 200°C and 400°C. During batch co-pyrolysis, the combination of the free radicals, especially stable and volatile radicals, lead to a reduction in both radical concentration and mass loss. In addition, the main temperature range of free radical interaction was 380°C ~ 600°C [32]. When the heating rate is slow enough, the pyrolysis of coal and biomass may occur independently from each other, and can be clearly distinguished. The synergistic effects could be limited, so it is easier to observe the additive effect. However, Wu et al. [33] observed that the thermogravimetric curve is not equal to the accumulation of the thermogravimetric

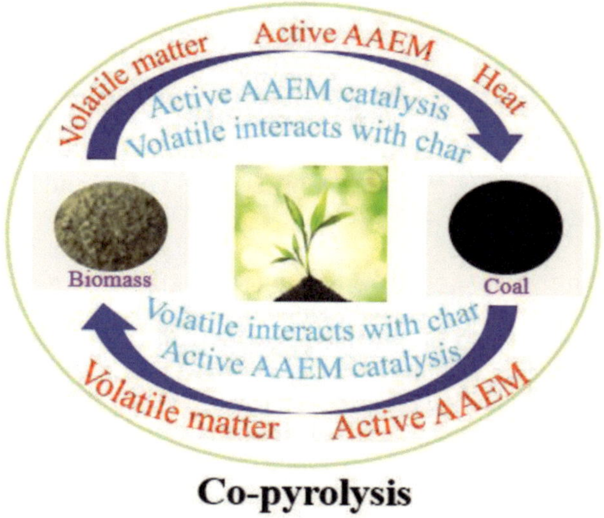

Figure 2.
Factors leading to synergies during co-pyrolysis of coal and biomass [31].

curve of the parent fuels, which indicated that there may be synergistic effects during co-pyrolysis. Even in the slow pyrolysis process, the migration of AAEMs can also occur. If the biomass is in close contact with the coal, the migration process is more obvious, and the catalytic effects of AAEMs on volatiles are more prominent. During slow co-pyrolysis of coal and biomass, when the biomass is in the main pyrolysis stage, coal is in the initial pyrolysis stage, and the volatiles generated by the biomass pyrolysis can interact with coal char. When coal is in the main pyrolysis stage, and biomass is in its secondary degassing stage, the volatiles produced by coal pyrolysis can also interact with biomass char [34]. Therefore, during the slow pyrolysis process, even if the main pyrolysis stages of coal and biomass are independent of each other, there are still synergistic or inhibitory effects.

During co-pyrolysis, when the heating rate is fast enough, the pyrolysis processes of coal and biomass can occur simultaneously, and the release of volatiles also overlap. The interaction between volatiles and co-pyrolysis char generated from coal and biomass can occur more easily through the following processes. The volatiles produced by biomass pyrolysis are rich in OH free radicals, H free radicals, and a small amount of other free radicals, which can move to the surface of the coal char, and enter into the char matrix [35]. A large number of fragment structures produced by depolymerization and decomposition of the coal matrix are combined with the above-mentioned small free radicals derived from biomass volatiles, which can inhibit the secondary cracking reaction [34]. In addition, AAEMs contained in biomass volatiles can be moved to the coal char, and AAEMs can significantly promote secondary cracking of volatiles generated from coal pyrolysis [36]. The volatiles produced by coal pyrolysis pass through the surface of the biomass char during the release process, and are catalyzed by AAEMs attached to the surface of biomass char to generate small molecular gases and macromolecular structures. The small molecular gases directly escape, due to their small steric hindrance, but the macromolecular structures remain in the biomass char matrix, and merge with it to form solid products [34]. From literature references since 2010, on the rapid co-pyrolysis of coal and biomass, it can be found that 83% of the studies reported that there are synergistic effects [11]. Yang et al. [37] researched the synergistic effect of cotton stalk (CS) and high-ash coal (HAC) on gas production during co-pyrolysis/gasification, and summarized the main reasons for the synergistic effect.

1. Higher pyrolysis temperature and narrow space were conducive to the diffusion of biomass-derived AAEMs during co-conversion. At high temperatures, the intermediate products from pyrolysis/gasification may participate in the reforming reaction, and produce non-condensable gases (H_2, CO, CO_2 and CH_4). As shown in **Figure 3**, changing the flow mode can prolong the contact time between volatiles and residual char (heterogeneous volatilization-char reaction), thereby increasing the yield of H_2 and CO, and reducing the yield of CO_2.

2. During the co-pyrolysis of coal and biomass, the biomass would be rapidly decomposed, and the biomass tar would be adsorbed onto the active sites of the residual char. As the proportion of cotton stalk in the mixture increases, the CO yield increased. This may be because AAEMs in the char promoted the decomposition of residual tar on the char (tar → H_2, CO; R(1) and R(2), see **Table 3**) [38]. In addition, the CO_2 generated inside the carbon matrix reacted with the char to expand the pores, while the light volatiles reacted with the char on the surface of the char, and consumed the carbon matrix (R(4)) [39]. Moreover, methane reforming (R(7)) and methane decomposition (R(8)) were performed simultaneously at high temperatures.

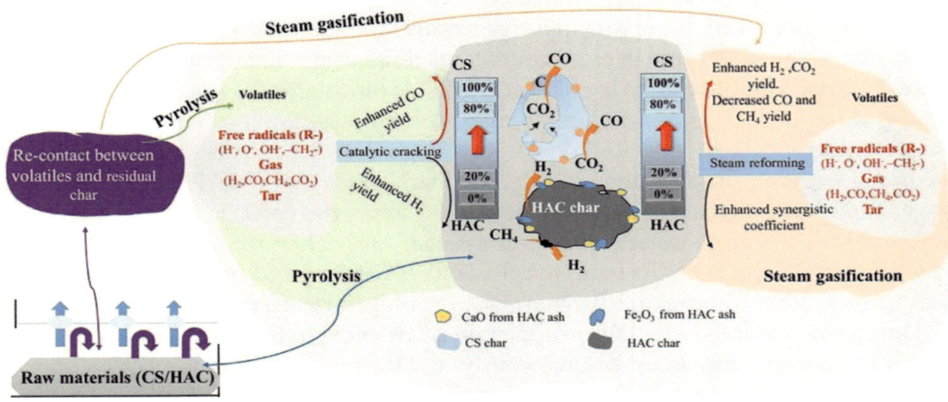

Figure 3.
Synergistic mechanisms and reactions in the co-pyrolysis/gasification of CS and HCS [37].

	Reactions	Number
First step	CS/HCS → Gas (H_2, CO, CO_2, CH_4, and others) + Tar (C_nH_m) + char	R(1)
Second step	Tar (C_nH_m) → Gas (H_2, CO, CO_2, CH_4, and others)	R(2)
	Char + H_2O → H_2 + CO; Char +2H_2O → 2H_2 + CO_2	R(3)
	Char + CO_2 → 2CO	R(4)
	CO + H_2O → H_2 + CO_2	R(5)
	CH_4 + H_2O → 3H_2 + CO; CH_4 + 2H_2O → 4H_2 + CO_2	R(6)
	CH_4 + CO_2 → 2CO + 2H_2	R(7)
	CH_4 → C + 2H_2	R(8)

Table 3.
Important reactions in the co-pyrolysis/gasification of CS and HCS [37].

3. The catalysis of AAEMs in cotton stalk can be enhanced by the addition of H_2O vapor to promote the reaction of active OH, H and – CH– free radicals with high-ash coal, thereby accelerating the C=C cleavage [28, 40]. During co-gasification, with the increase of cotton stalk ratio, the contents of H_2 and CO_2 increased, while the contents of CO and CH_4 decreased. This was due to the promoted heterogeneous carbon-vapor reactions (R(3) and R(4)) [41], water-gas shift reaction (R(5)), and homogeneous hydrocarbon reforming reactions (R(6) and R(7)) by inherent AAEMs [42].

4. Release and migration of AAEMs from biomass to coal during co-pyrolysis/co-gasification

AAEMs are easy to volatilize when participating in thermal conversion reactions of fuels [43]. In particular, alkali metals have been recognized for a long time to play a key role in the formation of deposits on the heat exchange surfaces of boilers in power stations [44–46]. In addition, the residual AAEMs in char could be effective catalysts for char reactions [43]. Therefore, understanding the release and migration behavior of AAEMs during gasification may assist in the development of AAEMs control technology, to effectively improve char reactivity.

AAEMs in coal and biomass can be divided into three forms: water-soluble state, ion-exchangeable state, and insoluble state. Water-soluble AAEMs and ion-exchangeable AAEMs are collectively called active AAEMs, while insoluble AAEMs are called inert AAEMs [31]. During the process of thermochemical conversion of fuel, active AAEMs can play a prominent catalytic role [14].

During the gasification process, the release of active AAEMs is not only due to evaporation to the gas phase, in the form of inorganic salts such as KCl and NaCl, but also due to release through substitution reactions. During high temperature pyrolysis, a large number of free radicals are generated, which can replace AAEMs bonded to the organic carboxyl group or other functional groups in the form of chemical bonds. It causes the chemical bonds between the char matrix and AAEMs to break, and AAEMs to be released. It is usually approximated by the following reaction [35]:

$$\mathrm{R} + CM - \mathrm{X} \rightarrow CM - \mathrm{R} + \mathrm{X} \tag{1}$$

where CM stands for the char matrix, X stands for AAEMs, and R stands for free radical. The valence of the element is another factor affecting the release of AAEMs. Generally, the alkali metals (Na and K) bonded with the functional groups by single bonds are more likely to be released than the alkaline earth metals (Ca and Mg) bonded by double bonds [28, 47]. In addition, the higher the pyrolysis temperature and heating rate, the more AAEMs are released from the coal and biomass [48–50].

However, the migration of AAEMs during co-gasification of coal and biomass is very different from that of coal or biomass gasification alone, and coal could be a key factor for AAEMs migration. Wei et al. [51] pointed out that the co-pyrolysis process mainly promoted the transfer of active K in the co-pyrolysis char, which weakened with the increase of biomass content in the mixture, but the transfer of active Ca was affected by the type of fuel. Ellis et al. [52] pointed out that the catalytically active calcium in biomass minerals and aluminosilicate minerals in coal can react to produce catalytically inert chabazite crystals during co-pyrolysis. Meng et al. [53] found that the content of AAEM in co-pyrolysis char increased with the increase of biomass ratio in the mixture, which was consistent with the conclusions obtained by Weiland et al. [54, 55]. Zhang et al. [10] indicated that K in biomass could be transferred to the surface of coal char during co-pyrolysis and co-gasification. Chen et al. [56] reported that the presence of coal during co-pyrolysis was not conductive to the volatilization of K and Mg in biomass, but the mixing of coal and corn stalks was conducive to the volatilization of Ca. Guanghui Hu [57] found that during co-pyrolysis of coal and biomass, the amount of K/Na in the biomass released into the gas phase was reduced, and the higher the pyrolysis temperature, the higher the content of K/Na in the char. Tao Ding [58] also found that the volatile amount of K/Na during the co-gasification process of coal and biomass was far less than that during the separate gasification process. Changchun Hu [59] believed that when the co-pyrolysis temperature of coal and biomass exceeded 460°C, the migration of K and Na would occur. In addition, during the subsequent gasification process, the K and Na migrated from the biomass to the coal char could combine with the minerals in coal and be fixed in the ash. It can be concluded that the coupled utilization of coal and biomass can alleviate the high temperature corrosion caused by the release of alkali metals when biomass is used alone. Lin et al. [60] concluded that co-pyrolysis under moderate temperature strongly favored inhibiting potassium from releasing, probably by interfering with free radical reactions. Song et al. [61] researched the migration path of K in biomass during thermal co-processing of coal and biomass (see **Figure 4**), and found that the mixed raw materials released 84.1 wt% (coal char 65.0 wt%, biochar 19,1 wt%) of biomass-K into the co-pyrolysis char, while only 15.9 wt% of biomass-K was released into the gas phase. The biomass-K migrated from the biomass to the coal char, and biochar was in the water-soluble (6.6 and 11.2 wt %, respectively), acetic acid-soluble (0.9 and 1.4 wt%, respectively), H_2SO_4-soluble (8.5 and 1.5 wt%, respectively), and H_2SO_4-insoluble (49.0 and 5.0 wt%, respectively) forms. After gasification, biomass-K accounted for 28.7% wt% in gas phase and 55.4 wt% in ash. Masnadi et al. [62] proposed four possible ways to lose active K during co-gasification: (1) volatilization; (2) forming inert alkali silicate; (3) forming new inert minerals (such as $KAlSiO_4$, $KAlSi_3O_8$) through irreversible reaction with minerals or ash in coal; (4) diffusion or implantation from the reaction surface into the carbon matrix.

During the co-utilization of coal and biomass, in addition to the migration of K from the biomass, the remaining AAEMs in coal and biomass also migrate, but their chemical forms and their migration pathways are not clear. In addition, the migration of AAEMs in coal and biomass to the surface of char and their distribution in

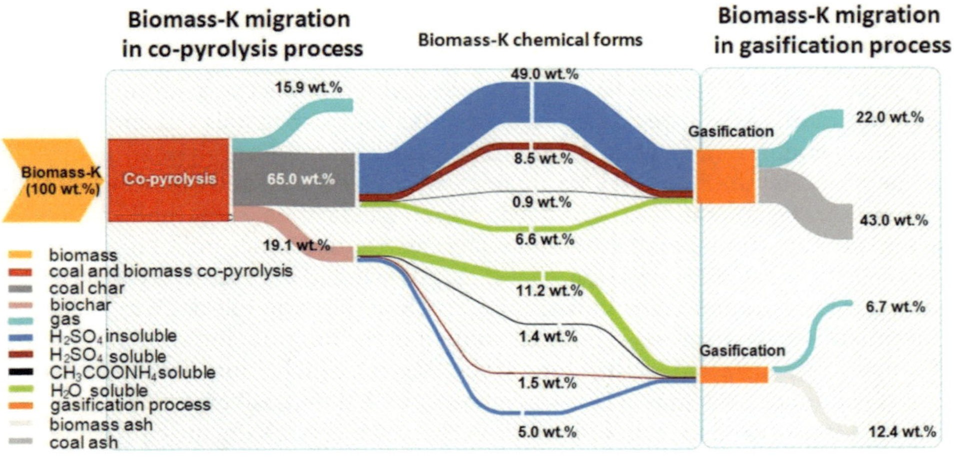

Figure 4.
Material flow of the biomass-K migration during thermal co-processing of coal and biomass. (pyrolysis: N_2 atmosphere, 1173 K; gasification: 1173 K, flow rate of H_2O/O_2 = 70/30 mL/min, gasification time = 20 min) [61].

solid phase and gas phase are not well understood. These problems could affect the efficiency of the co-utilization of coal and biomass. Therefore, it is necessary to conduct additional research on the chemical form, migration path, redistribution mechanism and evolution of AAEMs during co-gasification of coal and biomass.

5. Volatile-char interactions during co-pyrolysis of coal and biomass

The volatile-char interactions are common phenomena in the thermochemical conversion of low order fuels, and their interaction mechanisms are complex. The essence of volatile-char interactions is the reaction between char and H radicals produced by the cracking and reforming of volatiles. The interactions include not only the catalytic reforming effect of volatiles by char, but also the influence of volatiles on the structure and properties of char [63]. The volatile-char interactions can significantly affect many aspects of the gasification process, such as the volatilization of AAEMs, the evolution of char structure, the dispersion of inherent catalysts and thus the reactivity of char [64]. Therefore, the volatile-char interactions should be fully considered in the utilization of low-order fuels, the beneficial aspects of the volatile-char interactions should be strengthened while the adverse aspects should be weakened or eliminated (see **Figure 5**).

The volatile-char interactions during co-pyrolysis of coal and biomass are more complicated than that of each. Krerkkaiwan et al. [65, 66] found that coal char had catalytic effects on the decomposition of biomass volatiles and heavy aromatic hydrocarbons, and the interactions between biomass volatiles and coal char seriously reduced the gasification reactivity of coal char. Xia Wang [67] reported that lignin volatiles were more difficult than cellulose volatiles to undergo cracking and reforming reactions on the surface of coal char. When the gasification temperature was less than 800°C, biomass volatiles could form carbon deposits on the surface of coal char, which reduced the gasification reactivity of the coal char. Yan et al. [68] indicated that the interactions between biomass volatiles and coal char could reduce the yield of tar, increase the yield of gas volatiles, and change the chemical structure of the coal char. Hu et al. [69] found that the volatile-char interactions can promote the further cracking of tar into non-condensable gas, and can promote the aromatization of char, leading to the reduction of its gasification reaction. It can be found that the volatile-char interactions during co-pyrolysis of coal and biomass have important effects on the characteristics of both volatiles and co-pyrolysis char.

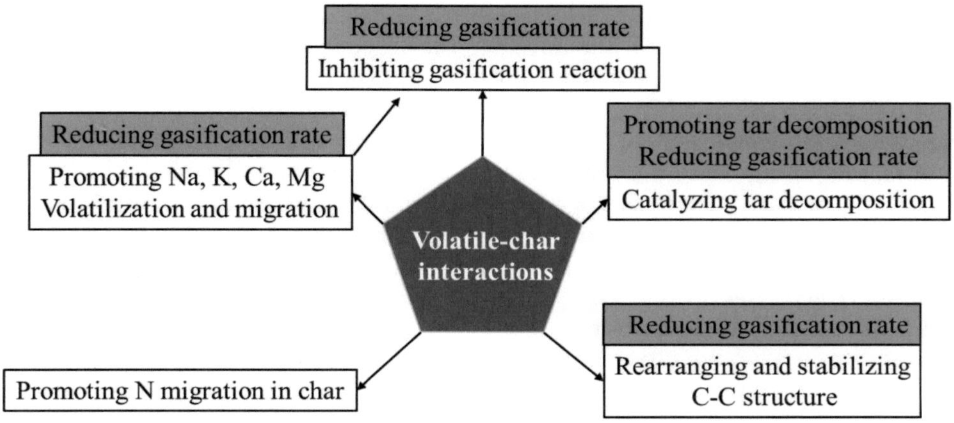

Figure 5.
The effect of volatile-char interactions on low rank gasification [63].

During thermochemical conversion of low-order fuel, the volatile-char interactions and AAEMs are interrelated and mutually influenced. On the one hand, the volatile-char interactions can promote the volatilization and migration of AAEMs. On the other hand, active AAEMs can affect the dynamic pyrolysis process of fuel, and have direct catalytic effects on the cracking of pyrolysis volatiles and their precursors, thus affecting the volatile-char interactions during thermochemical conversion. During co-thermochemical conversion of coal and biomass, the relationship between volatile-char interactions and AAEMs is more complex, and it is also a consideration for equipment design and operation. However, there are few reports on this aspect. Therefore, it is recommended that further research be conducted to understand the chemical mechanism of active AAEMs on volatile-char interactions and the influence of these interactions on the volatilization and migration of AAEMs for the efficient utilization of coal and biomass.

6. Co-pyrolysis products properties of coal and biomass

The pyrolysis products of coal and biomass include gas, tar, and char, which can be affected by synergistic effects. During co-pyrolysis of coal and biomass, the synergy can be affected by many factors such as: fuel type, blending ratio, heating rate, reactor design, and pyrolysis temperature. **Table 4** showed the effects of temperature and blending ratio on yields of char, tar and gas compared with calculated values. It can be found that the synergy shown by the pyrolysis product yield is not uniform, and can be affected by fuel type, pyrolysis temperature and blending ratio. In addition, the maceral group from low-rank coal can also affect the co-pyrolysis products. Wu et al. [74] researched the main maceral group from low-rank coal and cellulose in lignocellulosic biomass, and found that during co-pyrolysis, the influence of vitrinite on the generation of volatiles was related to the mixing ratio, while inertinite inhibited the generation of volatiles.

6.1 Co-pyrolysis volatiles composition of coal and biomass

Volatiles content is one of the important indicators of fuel characteristics, and it has important influences on furnace volume and shape, burner type and air distribution mode. At present, the research on volatiles during co-pyrolysis of coal and biomass mainly focuses on the influence of co-pyrolysis process on the composition and content of volatiles. Wu et al. [75, 76] found that when coal was co-pyrolyzed with wheat straw/biomass model compounds, co-pyrolysis promoted the generation of H_2 and CO, and inhibited the production of CO_2. However, the co-pyrolysis of coal and green algae inhibited the generation of H_2 and CO. Zhang et al. [77] reported that during co-pyrolysis of coal and biomass, the contents of volatile components (H_2, CO, CH_4 and CO_2) were inconsistent with the calculated values, suggesting that there were synergistic effects during co-pyrolysis. Sonobe et al. [30] indicated that the co-pyrolysis of coal and biomass had little effect on the production of CO and CO_2, but significantly promoted the production of CH_4. Soncini et al. [78] pointed out that the increase of biomass during the co-pyrolysis inhibited the production of CH_4, C_2H_4, CO and H_2. Yang et al. [37] researched the gas yield and the gas concentration of the co-pyrolysis of cotton stalk and high-ash coal at different mixing ratios under 950°C, and found that the co-pyrolysis was beneficial to the generation of gas, and can promote the formation of H_2, CO and CH_4 (except 20% cotton stalk), but inhibit the generation of CO_2. Wu et al. [79] reported that during co-pyrolysis of coal and biomass, the addition of low-rank coal inhibited the formation of CH_4 and H_2, and the negative synergistic effect was most

Pyrolysis condition				Co-pyrolysis products			#
Samples	Reactor	R_C (w/w)	T (°C)	Gas	Tar	Char	
				Y_E/Y_C (%)	Y_E/Y_C (%)	Y_E/Y_C (%)	
lignite, safflower seed	fixed-bed reactor	3%	550	24.1/24.7	36.7/33.4	22.0/23.3	[70]
		5%		23.5/24.4	39.6/32.9	21.6/24.0	
		7%		23.7/24.1	37.4/32.4	22.6/24.8	
		10%		23.6/23.6	35.6/31.7	24.4/26.0	
		20%		21.9/22.0	31.4/29.2	28.3/29.9	
		30%		19.8/20.3	25.4/26.6	34.5/33.8	
		50%		17.3/17.1	20.8/21.6	41.2/41.6	
		65%		14.4/19.5	16.8/25.4	46.9/35.7	
		90%		10.7/10.6	11.0/11.5	56.6/57.1	
sub-bituminous, sawdust	fixed-bed reactor	40%	400	21.2/15.2	41.0/45.5	37.8/39.2	
			500	21.4/16.3	43.3/47.4	35.2/36.3	[71]
			600	28.5/23.3	40.7/43.1	30.8/33.6	
			700	32.8/28.0	36.7/39.8	30.6/32.2	
			800	35.7/33.6	33.2/34.6	31.0/31.8	
		20%	600	27.4/26.3	50.1/50.1	22.1/23.6	
		40%		28.5/23.3	40.7/43.1	30.8/33.6	
		60%		26.3/20.2	33.2/36.2	40.5/43.6	
		80%		19.3/14.2	28.0/29.3	52.7/53.5	
low-rank coal, cedar	fixed-bed reactor	25%	450	18.0/18.2	33.6/35.4	34.8/34.2	[72]
			500	22.0/22.2	33.2/35.0	32.2/31.2	
			550	26.3/26.8	31.3/32.0	31.0/29.6	
			600	30.3/33.0	29.7/31.1	28.5/24.9	
			650	33.2/35.6	29.0/29.9	26.9/23.5	
		50%	450	16.5/15.2	28.0/30.1	46.4/45.2	
			500	20.5/18.7	26.9/30.1	43.5/41.7	
			550	25.1/23.0	25.8/27.2	40.1/39.6	
			600	28.7/28.2	25.6/26.9	38.0/35.6	
			650	31.5/30.6	25.2/26.0	36.8/34.1	
		75%	450	14.0/12.1	20.3/24.9	57.3/56.2	
			500	17.2/15.2	21.0/25.1	53.3/52.2	
			550	21.0/19.1	19.0/22.5	50.2/49.5	
			600	24.3/23.4	19.6/22.7	47.9/46.3	
			650	26.4/25.5	20.1/22.2	46.3/44.6	
Lignite, Pine sawdust	fixed-bed reactor	20%	400	41.9/31.6	27.5/30.1	30.6/38.1	[73]
			600	43.1/46.8	13.1/19.9	26.6/29.4	
			900	51.7/49.7	25.7/27.4	22.6/22.9	
		50%	400	35.0/27.2	15.5/20.3	49.6/52.5	
			600	43.5/37.5	15.4/18.9	41.2/43.5	
			900	52.8/44.0	13.1/19.9	34.1/36.1	
		80%	400	31.0/22.8	8.5/10.4	60.5/66.9	
			600	34.5/28.2	11.1/14.0	54.4/57.7	
			900	41.8/38.3	8.5/12.4	49.7/49.3	

R_C: mixing ratio of coal/mix (w/w); T: temperature; Y_E, Y_C: the experimental value and calculated value of yield, respectively.

Table 4.
Effects of temperature and blending ratio on yields of char, tar and gas compared with calculated values.

significant at a 50% mass ratio. Ma et al. [80] found that under the condition of cow manure: coal = 1: 3, CO emissions were significantly increased, CO_2 and CH_4 were also increased, and co-pyrolysis were beneficial to syngas production. In addition, for sulfur-containing gases, with the increase of cow manure ratio, the emissions of H_2S, COS and C_4H_4S increase, while the emission of SO2 decrease. Zhu et al. [72] pointed out that the synergistic effect of gas yield and composition during co-pyrolysis of coal and biomass was affected by pyrolysis temperature and mix ratio. However, to our best knowledge, there are no reports about the influence of AAEMs on the volatiles production characteristics of the co-pyrolysis of coal and biomass.

6.2 Co-pyrolysis tar properties of coal and biomass

The composition of tar is extremely complex, and can be used after separation and purification. The tar fractions are further processed to separate a variety of products. The main products extracted are: naphthalene, phenol, phenanthrene, carbazole, and asphalt. The different properties of biomass and coal lead to great differences in the components of their pyrolysis tar. Coal pyrolysis mainly produces heavy tar, while biomass pyrolysis mainly produces light tar. The interaction between coal and biomass during co-pyrolysis could cause changes in the properties of tar. Onay et al. [70] reported that the co-pyrolysis oil obtained with 5% lignite mixed with biomass contains more aliphatic and aromatic fractions, more relatively heavy hydrocarbons, less polar fractions than biomass pyrolysis oil. Jones et al. [81] found that co-pyrolysis of coal and biomass was conducive to the formation of phenols, but not conducive to the formation of aromatics. Tang et al. [82] indicated that co-pyrolysis improved the formation of phenols and naphthalene, while cotton stalk as an additive inhibited phenanthrene formation during co-pyrolysis of cotton stalk and Shenmu coal. Zhu et al. [72] concluded that the increase of cedar sawdust contributed to the positive synergistic effect of light tar, revealing the role of cedar as a hydrogen donor during co-pyrolysis. In addition, reactive H_2 from water-gas shift reaction and hydrogen-rich free radicals such as •CH_3, •OCH_3 from cedar can inhibit the secondary polymerization to form methyl-contained phenols and naphthalenes instead of 3-ring phenanthrenes and 4-ring pyrenes. Song et al. [73] reported that co-pyrolysis was unfavorable to the formation of benzene, naphthalene, and hydrocarbons in tar, but favorable to the formation of phenols and guaiacol. Zhao et al. [83] found that the co-pyrolysis of cellulose and lignite was conducive to the generation of −OH components, and cellulose could promote the thermal conversion of lignite to a certain extent, resulting in more ketones or esters in the co-pyrolysis tar, which was conducive to improving the quality of liquid products. Zhu et al. [84] reported that the reactive H_2 from water-gas shift reaction and hydrogen-rich radicals such as •CH_3, •OCH_3 from cedar can inhibit the secondary polymerization to form methylphenol and naphthalene instead of 3-ring phenanthrenes and 4-ring pyrenes during co-pyrolysis of a massive coal and cedar mixture.

6.3 Co-pyrolysis char properties of coal and biomass

The interaction between coal and biomass during co-pyrolysis can affect the characteristics of the resulting char, which influences the subsequent reactions. **Figure 6** shows the relationship between characteristics difference of coal and biomass and their co-pyrolysis char properties. Therefore, it is useful to study the characteristics of the resulting char, to understand the conversion mechanism during the co-pyrolysis of coal and biomass. Generally, the characteristics of char are studied from four aspects: physical structure, chemical structure, AAEM migration and reactivity.

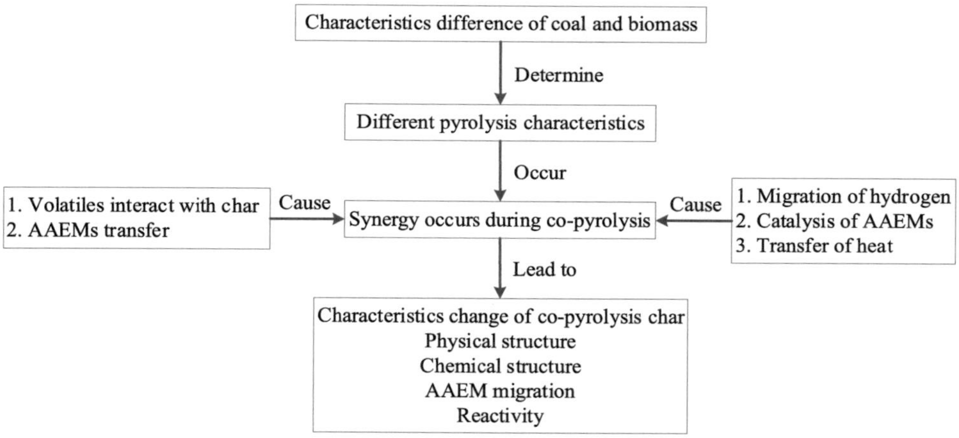

Figure 6.
The relationship between characteristics difference of coal and biomass and their co-pyrolysis char characteristics.

The change of the surface morphology of the co-pyrolysis char is the most direct manifestation of the interaction between coal and biomass. Wu et al. [76] found that with the increase of co-pyrolysis temperature, the surface of co-pyrolysis char became rougher and the pores became more developed. Wu et al. [85] reported that cellulose can promote the uniformity of co-pyrolysis char, while hemicellulose, lignin and sodium carboxymethyl cellulose were conducive to the three-dimensional development of co-pyrolysis char. Chen et al. [31] pointed out that during co-pyrolysis of coal and biomass, the presence of corn stalks was beneficial to the production of spherical particles from coal char. But the promotion of the corn stalks in removing active AAEMs through acid pickling was greater than that of the original corn stalks. However, Lin et al. [60] concluded that co-pyrolysis hardly influenced the macro-morphology, and structure of the mineral matter. According to current research results, more than 80% of the researches believed that the surface morphology of co-pyrolysis char was changed during co-pyrolysis [11].

During the reaction process of co-pyrolysis char, the pore structure can provide reaction and diffusion channels for some products, and the pore surface is the main location of chemical reaction of adsorption. Therefore, the structure and surface of pores in the char, play an important role in the reaction of chars. Most studies indicate that the growth of pore structure of co-pyrolysis char is affected by the operating temperature, mixing ratio and fuel type. Wei et al. [86] reported that with the increase of the biomass ratio in the mixture, the development of pore structure was first inhibited and then promoted. Wu et al. [87] found that biomass had an inhibitory effect on the development of the surface area and pore volume of co-pyrolysis char, but the average pore size of co-pyrolysis char was affected by the type of raw materials used. Vyas et al. [88] pointed out that the growth of co-pyrolysis char pore structure was affected by operating temperature and mixing ratio. At low co-pyrolysis temperature, the mixing ratio had little effect on the surface area and micropores. However, with the increase of co-pyrolysis temperature, the presence of biomass in the mixture significantly increased the number of micropores in the co-pyrolysis char. Lin et al. [60] found that specific surface area and pore structures of large micropores and mesopores were more impacted than those of ultramicropores, indicating that the influence was mainly from the formation of secondary char.

The chemical structure of organic compounds in co-pyrolysis char is also one of the important factors affecting the reactivity of the char. However, there is little research on the chemical structure of co-pyrolysis char, especially the functional group structure. Only a few researchers have studied the carbon structure of co-pyrolysis char by Raman spectroscopy. It was found that the more biomass in the mixture, the more favorable the formation of smaller (3–5 rings) aromatic ring structures, and the reduction of larger (no less than 6 rings) aromatic ring structures in the co-pyrolysis char [86]. Wu et al. [85] found that the addition of cellulose inhibited the formation of smaller aromatic ring structures in co-pyrolysis char, while the addition of hemicellulose and lignin contributed to the formation of smaller aromatic ring structures. Chen et al. [56] researched the effects of pyrolysis temperature on the structure and functional groups changes of co-pyrolysis char by Raman and Fourier transform Infrared spectroscopy (FTIR), respectively. They found that with the increase of pyrolysis temperature, the structure of co-pyrolysis char changed from a small aromatic ring system to a large aromatic ring system containing six or more condensed benzene rings through the condensation reaction of the rings, and the aromatic −CH functional groups first increased and then decreased. In addition, the C=O and aliphatic −CH functional groups in the co-pyrolysis char disappeared in the pyrolysis temperature from 600 to 700°C. Chen et al. [31] pointed out that active AAEMs can inhibit the decomposition of aliphatic −CH, C=O and − CH_3 in coal, and can also inhibit the decomposition of O − H, aliphatic −CH, and C − O in biomass. But as the pyrolysis temperature increased, the inhibitory effect gradually weakened or even dis appeared.

The co-gasification process of coal and biomass includes two primary steps: the co-pyrolysis of the raw fuel and the co-gasification of the remaining co-pyrolysis char. The gasification reaction rate of co-pyrolysis char is much slower than the release of volatiles during co-pyrolysis. Therefore, the reactivity of co-pyrolysis char becomes one of the important parameters, when evaluating the suitability of industrial gasification materials. It is still uncertain whether the co-pyrolysis process effects the reactivity of the co-pyrolysis char. Most studies found that no matter whether the gasification medium was CO_2, steam, or air, the co-pyrolysis process could affect the reactivity of the coal/biomass char. Some researchers believe that co-pyrolysis can inhibit the reactivity of the resulting char [52, 86, 89, 90], while others believe that it can promote the reactivity of char [10, 87, 91–95]. In addition, some researchers have found that the influence of co-pyrolysis on char reactivity was affected by gasification temperature [51, 96], mixing ratio [54, 97, 98], raw materials [54, 99] and other co-pyrolysis parameters [23]. Wei et al. [51] researched the co-gasification reactivity of rice straw and bituminous coal/anthracite mixed char; they found that the synergistic effect of co-gasification reactivity of rice straw and bituminous coal mixed char, gradually changed from inhibition to promotion. The co-gasification reactivity of rice straw-anthracite mixed char, gradually strengthened with the increase of the conversion rate, reached the strongest point in the middle stage of co-gasification process, and then began to slowly weaken. Chen et al. [23] also pointed out that with the progress of the co-gasification process, the synergistic effect of the gasification reactivity of the char gradually changed from inhibition to promotion. In addition, the gasification reactivity of co-pyrolysis char was affected by mixing ratio and co-pyrolysis temperature, which was consistent with the research conclusion of Yuan et al. [97], Gao et al. [94], and Mafu et al. [98]. Overall, according to current research results, more than 80% of the researches believed that there were synergistic effects of co-pyrolysis char during co-gasification [11].

The synergistic effect of co-gasification reactivity of co-pyrolysis char is mainly caused by two aspects: the interaction between coal and biomass during co-pyrolysis and the interaction between coal char and biomass char during co-gasification. However, existing research on the influence of these two aspects on the gasification reactivity of co-pyrolysis char is obviously insufficient. Only Chen et al. [23] compared the effects of these two aspects on the co-gasification reactivity; they found that the interaction between coal and biomass during co-pyrolysis had a more obvious impact on the co-gasification reactivity.

The gasification reactivity of co-pyrolysis char is most likely controlled by mass transfer, pore diffusion and internal chemical reaction. Therefore, for co-pyrolysis char of coal and biomass, its physicochemical structure and catalysis are the most important factors affecting its reactivity. A large number of studies have shown that the co-pyrolysis process can have a synergistic effect on the gasification reactivity of the resulting char, which was mainly due to the catalysis of AAEMs (mainly K and Ca) during the pyrolysis and gasification process [10, 51, 52, 92, 93, 99]. However, the catalytically active Ca and K in the biomass can interact with the aluminosilicate in coal minerals to form catalytically inert $Ca_2Al_2SiO_7$ and $KAlSiO_4$ crystals, thereby reducing the reactivity of co-pyrolysis char [22, 52, 99]. In addition, the rich silica components in biomass ash can also reduce the reactivity of chars by converting the catalytically active K and Ca substances into non-catalytically active substances [10]. Krerkkaiwan et al. [91] found that the reactivity of co-pyrolysis char was higher than that of coal char or biomass char, alone, which was related to the increased surface area and pore volume of co-pyrolysis char, as well as the catalytic effect of the K released by the biomass. Wei et al. [86] and Wang et al. [89] reported that the chemical structure of co-pyrolysis char and the migration of catalytically active AAEMs were the main factors affecting the reactivity of co-pyrolysis char, while the physical structure was a secondary factor. Wu et al. [87, 90] pointed out that the increase of the distance between microcrystalline structures and the number of interlayer defects between adjacent aromatic layers can promote the formation of active sites, thus increasing the reactivity of the co-pyrolysis char. Zhang et al. [100] found that the active AAEMs in coal can increase the reactivity of char during co-gasification, promote the production of H_2 and CO_2, and inhibit the production of CO. Chen et al. [23] found that the active AAEMs in biomass can obviously promote the reactivity of co-pyrolysis char, while the active AAEMs in coal had little effect on the reactivity of the co-pyrolysis char.

Current research on the co-pyrolysis products of coal and biomass seems to be limited to the macroscopic characteristics, such as the yield, properties, and compositions. There are few studies and analyses on the essential causes that affect the co-pyrolysis products. Therefore, it is impossible to clearly understand the production mechanism of products during co-pyrolysis of coal and biomass. In addition, the effect of active AAEMs on the volatiles production during co-pyrolysis of coal and biomass is still unclear, thus further research is recommended. A better understanding of the influence of the physicochemical structure and the active AAEMs on the reactivity of co-pyrolysis char, will be helpful to promote the development of industrial applications for biomass-coal co-pyrolysis. Although some researchers have begun to pay attention to this work, there are still plenty of opportunities for further research and development.

7. Industrial demonstration

Yao et al. [101] researched the industrial-scale co-pyrolysis of biomass, waste agriculture film, and bituminous coal, and analyzed it from multi-perspective (energy flow, economic, and socioenvironmental benefits analysis). The composition of different feedstock used in the pyrolysis experiment is shown in **Table 5**. The energy flow analysis showed that the co-pyrolysis processing of fruit tree branch (FTB), bituminous coal (BC), and recycled agriculture film pellets (AFP) resulted in a decrease in energy yield due to the energy loss that occurred during the conversion process. **Table 6** shows the results of the economic analysis of the industrial-scale

Feedstock type	Biomass	Fossil fuel	Plastic waste
Feedstock name	Fruit tree branch	Bituminous coal	Recycled agriculture film pellets
Feedstock code	FTB	BC	AFP
Experiment code	Relative contents added in each experiment (wt%)		
E1	100	0	0
E2	50	50	0
E3	40	40	20
Total mass processed (kg)[a]	450	800	1222

[a]*The total mass of material processed in each experiment was to keep the rotation speed constant based on their densities, and the total volume was fixed.*

Table 5.
The composition of different feedstock used in the pyrolysis experiment [101].

Category	Term	Unit	E1	E2	E3
Financial data	Total investment	10,000 CNY	213.79	223.37	223.37
	Fixed asset investment		204	204	204
	Average income during operation period		87.6	152.58	166
	Total cost (average during operation period)		72	129.5	129.5
	Total profit (average during operation period)		15.6	23.08	36.5
Financial evaluation indices	Financial internal rate of return (Before tax)	%	10.51%	14.20%	21.26%
	Financial net present value (Before tax)	10,000 CNY	31.09	82.99	185.44
	Payback period (Before tax)	year	8.47	7.19	5.51
Sensitivity analysis	Product critical point	%	4.73	7.05	14.66
	Feedstock critical point	%	10.14	11.07	25.66
	Initial investment critical point	%	16.29	37.69	64.65
BEP analysis	BEP (%, capacity utilization rate)	%	62.63%	53.11%	41.73%

Table 6.
Input data and results of the project economic analysis [101].

co-pyrolysis of FTB, BC, and AFP. From the economic analysis, it can be concluded that the three pyrolysis methods can bring economic benefits. Among them, the economic performance of FTB-BC-AFP co-pyrolysis was the highest, while that of FTB single pyrolysis was the lowest. The annual profit and the internal financial return rate of FTB-BC-AFP co-pyrolysis were three times and 2.1 times higher than that of single pyrolysis, respectively. In addition, the payback period can be shortened by about 3 years. The biochar produced by the three pyrolysis methods conformed to the national standard (GB/T 31862–2015 and GB/T 34170–2017).

The pyrolysis gas meted the calorific value requirements of the national standard (GB/T 13612–2006), and can meet the needs of residents for heating and cooking. Moreover, the implementation of the project has created employment opportunities, and each person can increase their income by 30,000 CNY per year. In addition, The FTB-BC-AFP co-pyrolysis used in this project can replace ~1100 tons of standard coal every year, and reduce CO2 emission, SO2 emission, smoke and other pollutants by 1720 tons, 5 ~ 6 tons, and 320 kg per annum, respectively. At the same time, the project recycled 750 tons of plastic waste, which can reduce 50–66.7 km^2 of farmland white pollution and avoid the accumulation of plastic waste.

8. Vision and development

In response to China's dual-carbon target, the use of coal should be reduced and eliminated, as burning 1 kg of coal produces 2.62 kg of CO_2. As a solid fuel with zero carbon emission, the utilization rate of biomass should be increased. China produces more than 1 billion tons of agriculture and forestry waste each year. Due to its low energy density and high transportation cost, on-site treatment of biomass can effectively reduce the cost of recycling. Although co-pyrolysis may lead to energy loss in the conversion process, the addition of coal increases the bulk density of the raw material mixture and improves the processing capacity of the equipment. In addition, co-pyrolysis can improve the combustion characteristics of char and reduce the emission of pollutants. Co-pyrolysis process can not only effectively meet the needs of clean energy in rural areas, but also realize the on-site treatment and utilization of these major solid wastes. Therefore, the co-pyrolysis of biomass and coal is still a valuable method for engineering applications that require the use of coal. Furthermore, the addition of plastics in co-pyrolysis can improve the yield and quality of gas products, and also has certain environmental benefits.

In summary, there are still many deficiencies in the current research on the co-pyrolysis of coal and biomass, and many opportunities to expand the knowledge of the resulting chemistry, e.g., the influence of the interaction between coal and biomass on the respective pyrolysis process and the entire co-pyrolysis process during co-pyrolysis, the influence of co-pyrolysis conditions on the physicochemical structure and AAEMs content of co-pyrolysis char, especially the influence of co-pyrolysis process on subsequent gasification reaction characteristics. Important system parameters to be studied further, include: (a) the ratio, and limits, of biomass to coal; (b) pyrolysis and gasification operating temperatures; (c) the rate of temperature rise in the reaction vessel; (d) the inter-catalytic effects of AAEMs on product yields and compositions; (e) the yields of volatiles (gases), tars (liquids), and chars (solids); (f) the compositions of these products. The goal of all of this current and future biomass-coal co-pyrolysis work should be to reach industrial scale applications for this as soon as possible. With regard to the current global climate change crisis, it is urgent to continue to minimize the use of all fossil fuels, worldwide, especially coal, and to mitigate the emissions of CO_2 into Earth's atmosphere. With that goal in mind, the growing use of biomass, to replace the use of coal, is of paramount importance.

9. Conclusion

This chapter has reviewed some of the information regarding the co-pyrolysis of coal and biomass, with a focus on the synergistic mechanism and the resulting influence. The different characteristics of coal and biomass lead to great differences in their pyrolysis characteristics, resulting in a synergistic effect during co-pyrolysis. The synergistic effect can be caused by the migration of active H radicals from biomass to coal, the catalysis of active AAEMs, and heat transfer during co-pyrolysis [102]. During the co-pyrolysis of coal and biomass, changes in product yields and composition of volatiles, as well as the changes in the physicochemical structure and reactivity of co-pyrolysis char are briefly reviewed. In addition, the release and migration of AAEMs and their catalytic effects, and volatile-char interactions are mentioned. Moreover, the analysis of the co-pyrolysis industry demonstration is also mentioned.

Acknowledgements

The authors gratefully acknowledge the China Postdoctoral Science Foundation Funded Project (2021M690824).

Author details

Xiye Chen[1,2], Linyao Zhang[1,2]*, Li Liu[1], Chang Xing[3], Yan Zhao[4], Kirk Cobb[5], Roger Ruan[5] and Penghua Qiu[1]*

1 School of Energy Science and Engineering, Harbin Institute of Technology, Harbin, China

2 School of Chemistry and Chemical Engineering, Harbin Institute of Technology, Harbin, China

3 School of Astronautics, Harbin Institute of Technology, Harbin, China

4 Liaoning Provincial Key Laboratory for Urban Ecology, Shenyang Academy of Environmental Science, Shenyang, China

5 Center for Biorefining and Department of Bioproducts and Biosystems Engineering, University of Minnesota, St. Paul, Minnesota, United States

*Address all correspondence to: zhanglinyao1209@gmail.com and qiuph@hit.edu.cn

References

[1] Mallick D, Mahanta P, Moholkar V S. Synergistic Effects in Gasification of Coal/Biomass Blends: Analysis and Review[C]//De S, Agarwal A K, Moholkar V S, et al. Coal and Biomass Gasification: Recent Advances and Future Challenges. Singapore: Springer Singapore, 2018:473-497.

[2] Balat M. Influence of Coal as an Energy Source on Environmental Pollution[J]. Energy Sources, Part A: Recovery, Utilization, and Environmental Effects, 2007, 29(7): 581-589.

[3] Saxena R C, Adhikari D K, Goyal H B. Biomass-based energy fuel through biochemical routes: A review[J]. Renewable and Sustainable Energy Reviews, 2009, 13(1): 167-178.

[4] Bridgewater A V. Biomass fast pyrolysis[J]. Thermal science, 2004, 8(2): 21-50.

[5] Chen X, Liu L, Zhang L, et al. Thermogravimetric analysis and kinetics of the co-pyrolysis of coal blends with corn stalks[J]. Thermochimica Acta, 2018, 659(Supplement C): 59-65.

[6] McKendry P. Energy production from biomass (Part 1): Overview of biomass[J]. Bioresour Technol, 2002, 83(1): 37-46.

[7] Ren X, Feng X, Cao J, et al. Catalytic Conversion of Coal and Biomass Volatiles: A Review[J]. Energy & Fuels, 2020, 34(9): 10307-10363.

[8] Xi J. Xi Jinping's speech at the general debate of the 75th UN General Assembly [EB/OL]. http://www.xinhuanet.com/politics/leaders/2020-09/22/c_112652764 7.htm.

[9] Report of China Biomass Energy Utilization Industry In-depth Analysis and Development Planning Advisory Recommendation (2020-2024)China Investment Industry Research Institute, 2020:

[10] Zhang Y, Zheng Y, Yang M, et al. Effect of fuel origin on synergy during co-gasification of biomass and coal in CO2[J]. Bioresource Technology, 2016, 200: 789-794.

[11] Chen X, Liu L, Zhang L, et al. A Review on the Properties of Copyrolysis Char from Coal Blended with Biomass[J]. Energy & Fuels, 2020, 34(4): 3996-4005.

[12] Xu C, Hu S, Xiang J, et al. Interaction and kinetic analysis for coal and biomass co-gasification by TG–FTIR[J]. Bioresource Technology, 2014, 154: 313-321.

[13] Habibi R, Kopyscinski J, Masnadi M S, et al. Co-gasification of Biomass and Non-biomass Feedstocks: Synergistic and Inhibition Effects of Switchgrass Mixed with Sub-bituminous Coal and Fluid Coke During CO2 Gasification[J]. Energy & Fuels, 2012, 27(1): 494-500.

[14] Li C. Advances in the science of Victorian brown coal[M]. Oxford: Elsevier, 2004.

[15] Verma M, Loha C, Sinha A N, et al. Drying of biomass for utilising in co-firing with coal and its impact on environment – A review[J]. Renewable and Sustainable Energy Reviews, 2017, 71: 732-741.

[16] Balasubramanian N, Srinivasakannan C. Drying of granular materials in circulating fluidized beds[J]. Advanced Powder Technology, 2007, 18(2): 135-142.

[17] Li C Z, Sathe C, Kershaw J R, et al. Fates and roles of alkali and alkaline earth metals during the pyrolysis of a

Victorian brown coal[J]. Fuel, 2000, 79(3-4): 427-438.

[18] Khan A A, de Jong W, Jansens P J, et al. Biomass combustion in fluidized bed boilers: Potential problems and remedies[J]. Fuel Processing Technology, 2009, 90(1): 21-50.

[19] Agbor E, Zhang X, Kumar A. A review of biomass co-firing in North America[J]. Renewable and Sustainable Energy Reviews, 2014, 40: 930-943.

[20] Bhuiyan A A, Blicblau A S, Islam A K M S, et al. A review on thermo-chemical characteristics of coal/biomass co-firing in industrial furnace[J]. Journal of the Energy Institute, 2018, 91(1): 1-18.

[21] Scarlat N, Dallemand J, Skjelhaugen O J, et al. An overview of the biomass resource potential of Norway for bioenergy use[J]. Renewable and Sustainable Energy Reviews, 2011, 15(7): 3388-3398.

[22] Mallick D, Mahanta P, Moholkar V S. Co-gasification of coal and biomass blends: Chemistry and engineering[J]. Fuel, 2017, 204: 106-128.

[23] Chen X, Liu L, Zhang L, et al. Gasification reactivity of co-pyrolysis char from coal blended with corn stalks[J]. Bioresource Technology, 2019, 279: 243-251.

[24] Vassilev S V, Baxter D, Andersen L K, et al. An overview of the chemical composition of biomass[J]. Fuel, 2010, 89(5): 913-933.

[25] Emami-Taba L, Irfan M F, Wan Daud W M A, et al. Fuel blending effects on the co-gasification of coal and biomass – A review[J]. Biomass and Bioenergy, 2013, 57: 249-263.

[26] Emami-Taba L, Irfan M F, Wan Daud W M A, et al. Fuel blending effects on the co-gasification of coal and biomass – A review[J]. Biomass and Bioenergy, 2013, 57: 249-263.

[27] Li S, Chen X, Liu A, et al. Study on co-pyrolysis characteristics of rice straw and Shenfu bituminous coal blends in a fixed bed reactor[J]. Bioresource Technology, 2014, 155: 252-257.

[28] Long J, Song H, Jun X, et al. Release characteristics of alkali and alkaline earth metallic species during biomass pyrolysis and steam gasification process[J]. Bioresource Technology, 2012, 116: 278-284.

[29] Ulloa C A, Gordon A L, García X A. Thermogravimetric study of interactions in the pyrolysis of blends of coal with radiata pine sawdust[J]. Fuel Processing Technology, 2009, 90(4): 583-590.

[30] Sonobe T, Worasuwannarak N, Pipatmanomai S. Synergies in co-pyrolysis of Thai lignite and corncob[J]. Fuel Processing Technology, 2008, 89(12): 1371-1378.

[31] Chen X, Liu L, Zhang L, et al. Effect of active alkali and alkaline earth metals on physicochemical properties and gasification reactivity of co-pyrolysis char from coal blended with corn stalks[J]. Renewable Energy, 2021, 171: 1213-1223.

[32] He W, Yin G, Zhao Y, et al. Interactions between free radicals during co-pyrolysis of lignite and biomass[J]. Fuel, 2021, 302: 121098.

[33] Wu Z, Yang W, Tian X, et al. Synergistic effects from co-pyrolysis of low-rank coal and model components of microalgae biomass[J]. Energy Conversion and Management, 2017, 135: 212-225.

[34] Chen X, Liu L, Zhang L, et al. Pyrolysis Characteristics and Kinetics of Coal–Biomass Blends during Co-Pyrolysis[J]. Energy & Fuels, 2019, 33(2): 1267-1278.

[35] Wu H, Quyn D M, Li C. Volatilisation and catalytic effects of alkali and alkaline earth metallic species during the pyrolysis and gasification of Victorian brown coal. Part III. The importance of the interactions between volatiles and char at high temperature[J]. Fuel, 2002, 81(8): 1033-1039.

[36] Quyn D M, Wu H, Hayashi J, et al. Volatilisation and catalytic effects of alkali and alkaline earth metallic species during the pyrolysis and gasification of Victorian brown coal. Part IV. Catalytic effects of NaCl and ion-exchangeable Na in coal on char reactivity☆[J]. Fuel, 2003, 82(5): 587-593.

[37] Yang P, Zhao S, Zhang Q, et al. Synergistic effect of the cotton stalk and high-ash coal on gas production during co-pyrolysis/gasification[J]. Bioresource Technology, 2021, 336: 125336.

[38] Kern S, Pfeifer C, Hofbauer H. Gasification of lignite in a dual fluidized bed gasifier — Influence of bed material particle size and the amount of steam[J]. Fuel Processing Technology, 2013, 111: 1-13.

[39] Li L, Tong S, Duan L, et al. Effect of CO2 and H2O on lignite char structure and reactivity in a fluidized bed reactor[J]. Fuel Processing Technology, 2021, 211: 106564.

[40] Alvarez J, Lopez G, Amutio M, et al. Evolution of biomass char features and their role in the reactivity during steam gasification in a conical spouted bed reactor[J]. Energy Conversion and Management, 2019, 181: 214-222.

[41] Hu J, Shao J, Yang H, et al. Co-gasification of coal and biomass: Synergy, characterization and reactivity of the residual char[J]. Bioresource Technology, 2017, 244: 1-7.

[42] Yu J, Guo Q, Gong Y, et al. A review of the effects of alkali and alkaline earth metal species on biomass gasification[J]. Fuel Processing Technology, 2021, 214: 106723.

[43] Li C. Some recent advances in the understanding of the pyrolysis and gasification behaviour of Victorian brown coal[J]. Fuel, 2007, 86(12-13): 1664-1683.

[44] Wibberley L J, Wall T F. Alkali-ash reactions and deposit formation in pulverized-coal-fired boilers: experimental aspects of sodium silicate formation and the formation of deposits[J]. Fuel, 1982, 61(1): 93-99.

[45] Wibberley L J, Wall T F. Alkali-ash reactions and deposit formation in pulverized-coal-fired boilers: the thermodynamic aspects involving silica, sodium, sulphur and chlorine[J]. Fuel, 1982, 61(1): 87-92.

[46] Boow J. Sodium/ash reactions in the formation of fireside deposits in pulverized-fuel-fired boilers[J]. Fuel, 1972, 51(3): 170-173.

[47] Quyn D M, Wu H, Li C. Volatilisation and catalytic effects of alkali and alkaline earth metallic species during the pyrolysis and gasification of Victorian brown coal. Part I. Volatilisation of Na and Cl from a set of NaCl-loaded samples[J]. Fuel, 2002, 81(2): 143-149.

[48] Okuno T, Sonoyama N, Hayashi J, et al. Primary Release of Alkali and Alkaline Earth Metallic Species during the Pyrolysis of Pulverized Biomass[J]. Energy & Fuels, 2005, 19(5): 2164-2171.

[49] Keown D M, Favas G, Hayashi J, et al. Volatilisation of alkali and alkaline earth metallic species during the pyrolysis of biomass: differences between sugar cane bagasse and cane trash[J]. Bioresource Technology, 2005, 96(14): 1570-1577.

[50] Zhao Y. Studies on the influence and volatilization behavior of alkali and

alkaline earth metals during pyrolysis of zhundong coal[D]. Harbin: Harbin Institute of Technology, 2017:

[51] Wei J, Gong Y, Guo Q, et al. A mechanism investigation of synergy behaviour variations during blended char co-gasification of biomass and different rank coals[J]. Renewable Energy, 2019, 131: 597-605.

[52] Ellis N, Masnadi M S, Roberts D G, et al. Mineral matter interactions during co-pyrolysis of coal and biomass and their impact on intrinsic char co-gasification reactivity[J]. Chemical Engineering Journal, 2015, 279: 402-408.

[53] Meng H, Wang S, Chen L, et al. Thermal behavior and the evolution of char structure during co-pyrolysis of platanus wood blends with different rank coals from northern China[J]. Fuel, 2015, 158: 602-611.

[54] Weiland N T, Means N C, Morreale B D. Product distributions from isothermal co-pyrolysis of coal and biomass[J]. Fuel, 2012, 94: 563-570.

[55] Wu Z, Wang S, Zhao J, et al. Thermal Behavior and Char Structure Evolution of Bituminous Coal Blends with Edible Fungi Residue during Co-Pyrolysis[J]. Energy & Fuels, 2014, 28(3): 1792-1801.

[56] Chen X, Liu L, Zhang L, et al. Physicochemical Properties and AAEM Retention of Copyrolysis Char from Coal Blended with Corn Stalks[J]. Energy & Fuels, 2019, 33(11): 11082-11091.

[57] Hu G. The Migration and Catalytic Characteristics of Alkali Metals during Coal and Biomass Co-pyrolysis Process[D]. Taiyuan: Taiyuan University of Technology, 2014:

[58] Ding T. The Migration Characteristics and Catalysis Contribution of Alkali Matals During Coal and Biomass Co-gasification[D]. Taiyuan: Taiyuan University of Technology, 2012:

[59] Hu C. The Migration and Catalytic Characteristics of Biomass Alkalis During Co-gasification of Coal and Biomass[D]. Taiyuan: Taiyuan University of Technology, 2015:

[60] Lin B, Zhou J, Qin Q, et al. Physicochemical characteristics of biomass-coal blend char: The role of co-pyrolysis synergy[J]. Energy Science & Engineering, 2021.

[61] Song Y, Li Q, Li F, et al. Pathway of biomass-potassium migration in co-gasification of coal and biomass[J]. Fuel, 2019, 239: 365-372.

[62] Masnadi M S, Grace J R, Bi X T, et al. From fossil fuels towards renewables: Inhibitory and catalytic effects on carbon thermochemical conversion during co-gasification of biomass with fossil fuels[J]. Applied Energy, 2015, 140: 196-209.

[63] Yang H, Li C, Zhang H, et al. Basic Principles of the Interaction Between Volatiles and Char in Thermochemical Convertion Process of Low Rank Carbonaceous Fuels[J]. Contemporary Chemical Industry, 2016, 45(09): 2252-2255.

[64] Li C. Importance of volatile–char interactions during the pyrolysis and gasification of low-rank fuels – A review[J]. Fuel, 2013, 112: 609-623.

[65] Krerkkaiwan S, Tsutsumi A, Kuchonthara P. Biomass derived tar decomposition over coal char bed[J]. ScienceAsia, 2013, 39(5): 511.

[66] Krerkkaiwan S, Fushimi C, Yamamoto H, et al. Influences of heating rate during coal char preparation and AAEMs on volatile–char interaction with different sources of biomass volatile[J]. Fuel Processing Technology, 2014, 119: 10-18.

[67] Wang X. Interactions between Biomass Volatiles and Coal Char during Coal and Biomass Co-gasification[D]. Taiyuan: Taiyuan University of Technology, 2016:

[68] Yan D, Guo J, Wang M, et al. Effects of interactions between corncob volatiles and lignite on their products during rapid co-pyrolysis[J]. Journal of Analytical and Applied Pyrolysis, 2019, 137: 212-219.

[69] Hu J, Si Y, Yang H, et al. Influence of volatiles-char interactions between coal and biomass on the volatiles released, resulting char structure and reactivity during co-pyrolysis[J]. Energy Conversion and Management, 2017, 152: 229-238.

[70] Onay Ö, Bayram E, Koçkar Ö M. Copyrolysis of Seyitömer−Lignite and Safflower Seed: Influence of the Blending Ratio and Pyrolysis Temperature on Product Yields and Oil Characterization[J]. Energy & Fuels, 2007, 21(5): 3049-3056.

[71] Park D K, Kim S D, Lee S H, et al. Co-pyrolysis characteristics of sawdust and coal blend in TGA and a fixed bed reactor[J]. Bioresource Technology, 2010, 101(15): 6151-6156.

[72] Zhu J, Jin L, Luo Y, et al. Fast co-pyrolysis of a massive Naomaohu coal and cedar mixture using rapid infrared heating[J]. Energy Conversion and Management, 2020, 205: 112442.

[73] Song Y, Tahmasebi A, Yu J. Co-pyrolysis of pine sawdust and lignite in a thermogravimetric analyzer and a fixed-bed reactor[J]. Bioresource Technology, 2014, 174: 204-211.

[74] Wu Z, Zhang J, Fan Y, et al. Synergistic effects from co-pyrolysis of lignocellulosic biomass with low-rank coal: A perspective based on the interaction of organic components[J]. Fuel, 2021, 306: 121648.

[75] Wu Z, Li Y, Xu D, et al. Co-pyrolysis of lignocellulosic biomass with low-quality coal: Optimal design and synergistic effect from gaseous products distribution[J]. Fuel, 2019, 236: 43-54.

[76] Wu Z, Yang W, Li Y, et al. Co-pyrolysis behavior of microalgae biomass and low-quality coal: Products distributions, char-surface morphology, and synergistic effects[J]. Bioresource Technology, 2018, 255: 238-245.

[77] Zhang L, Xu S, Zhao W, et al. Co-pyrolysis of biomass and coal in a free fall reactor[J]. Fuel, 2007, 86(3): 353-359.

[78] Soncini R M, Means N C, Weiland N T. Co-pyrolysis of low rank coals and biomass: Product distributions[J]. Fuel, 2013, 112: 74-82.

[79] Wu Z, Zhang J, Zhang B, et al. Synergistic effects from co-pyrolysis of lignocellulosic biomass main component with low-rank coal: Online and offline analysis on products distribution and kinetic characteristics[J]. Applied Energy, 2020, 276: 115461.

[80] Ma M, Bai Y, Song X, et al. Investigation into the co-pyrolysis behaviors of cow manure and coal blending by TG–MS[J]. Science of The Total Environment, 2020, 728: 138828.

[81] Jones J M, Kubacki M, Kubica K, et al. Devolatilisation characteristics of coal and biomass blends[J]. Journal of Analytical and Applied Pyrolysis, 2005, 74(1-2): 502-511.

[82] Tang C, Zhao J, Li X, et al. Formation Mechanism of Aromatics During Co-pyrolysis of Coal and Cotton Stalk[J]. BioResources, 2020, 2(15): 4449-4463.

[83] Zhao Y, Cao H, Yao C, et al. Synergistic effects on cellulose and lignite co-pyrolysis and co-liquefaction[J]. Bioresource Technology, 2020, 299: 122627.

[84] Zhu J, Jin L, Luo Y, et al. Fast co-pyrolysis of a massive Naomaohu coal and cedar mixture using rapid infrared heating[J]. Energy Conversion and Management, 2020, 205: 112442.

[85] Wu Z, Yang W, Chen L, et al. Morphology and microstructure of co-pyrolysis char from bituminous coal blended with lignocellulosic biomass: Effects of cellulose, hemicellulose and lignin[J]. Applied Thermal Engineering, 2017, 116: 24-32.

[86] Wei J, Gong Y, Guo Q, et al. Physicochemical evolution during rice straw and coal co-pyrolysis and its effect on co-gasification reactivity[J]. Bioresource Technology, 2017, 227: 345-352.

[87] Wu Z, Ma C, Jiang Z, et al. Structure evolution and gasification characteristic analysis on co-pyrolysis char from lignocellulosic biomass and two ranks of coal: Effect of wheat straw[J]. Fuel, 2019, 239: 180-190.

[88] Vyas A, Chellappa T, Goldfarb J L. Porosity development and reactivity changes of coal–biomass blends during co-pyrolysis at various temperatures[J]. Journal of Analytical and Applied Pyrolysis, 2017, 124: 79-88.

[89] Wang M, Tian J, Roberts D G, et al. Interactions between corncob and lignite during temperature-programmed co-pyrolysis[J]. Fuel, 2015, 142: 102-108.

[90] Wu Z, Wang S, Luo Z, et al. Physico-chemical properties and gasification reactivity of co-pyrolysis char from different rank of coal blended with lignocellulosic biomass: Effects of the cellulose[J]. Bioresource Technology, 2017, 235: 256-264.

[91] Krerkkaiwan S, Fushimi C, Tsutsumi A, et al. Synergetic effect during co-pyrolysis/gasification of biomass and sub-bituminous coal[J]. Fuel Processing Technology, 2013, 115: 11-18.

[92] Zhu W, Song W, Lin W. Catalytic gasification of char from co-pyrolysis of coal and biomass[J]. Fuel Processing Technology, 2008, 89(9): 890-896.

[93] Jeong H J, Park S S, Hwang J. Co-gasification of coal-biomass blended char with CO2 at temperatures of 900-1100°C[J]. Fuel, 2014, 116: 465-470.

[94] Gao C, Vejahati F, Katalambula H, et al. Co-gasification of Biomass with Coal and Oil Sand Coke in a Drop Tube Furnace†[J]. Energy & Fuels, 2010, 24(1): 232-240.

[95] Jeong H J, Hwang I S, Hwang J. Co-gasification of bituminous coal–pine sawdust blended char with H 2 O at temperatures of 750-850 °C[J]. Fuel, 2015, 156: 26-29.

[96] Jiao Z, Liu L, Zhao Y, et al. Study on Reactivity and Synergy Behavior of Cogasification between Biomass Char and Coal Char[J]. Energy & Fuels, 2021, 35(1): 341-350.

[97] Yuan S, Dai Z, Zhou Z, et al. Rapid co-pyrolysis of rice straw and a bituminous coal in a high-frequency furnace and gasification of the residual char[J]. Bioresource Technology, 2012, 109: 188-197.

[98] Mafu L D, Neomagus H W J P, Everson R C, et al. The carbon dioxide gasification characteristics of biomass char samples and their effect on coal gasification reactivity during co-gasification[J]. Bioresource Technology, 2018, 258: 70-78.

[99] Ding L, Zhang Y, Wang Z, et al. Interaction and its induced inhibiting or synergistic effects during co-gasification of coal char and biomass char[J]. Bioresource Technology, 2014, 173: 11-20.

[100] Zhang Z, Pang S, Levi T. Influence of AAEM species in coal and biomass on steam co-gasification of chars of blended coal and biomass[J]. Renewable Energy, 2017, 101: 356-363.

[101] Yao Z, Kang K, Cong H, et al. Demonstration and multi-perspective analysis of industrial-scale co-pyrolysis of biomass, waste agricultural film, and bituminous coal[J]. Journal of Cleaner Production, 2021, 290: 125819.

[102] Quan C, Gao N. Copyrolysis of Biomass and Coal: A Review of Effects of Copyrolysis Parameters, Product Properties, and Synergistic Mechanisms[J]. BioMed Research International, 2016, 2016: 1-11.

Chapter 5

Algae Based Bio-Plastics: Future of Green Economy

Arathi Sreenikethanam and Amit Bajhaiya

Abstract

Plastic has become one of the most crucial requirements of the modern-day living. The continuous reliance on the petroleum-based, non-biodegradable plastics has resulted in increased global environmental damage and rapid depletion of fossil fuels. Bioplastic, with remarkably similar properties to petroleum-based plastics is a promising alternative to overcome these emerging challenges. Despite the fact that algae and cyanobacteria are feasible alternative source for bio-plastic, there have been limited studies on strain selection and optimization of culture conditions for the bio plastic production. Naturally, algae and cynobacteria can accumulate higher amount of metabolites under stress conditions however one of the recent study on genetic engineering of *Synechocystis sp.* coupled with abiotic stresses showed up to 81% of increase in PHB level in the transformed lines. This chapter provides summary of various studies done in the field of algal bio-plastics, including bioplastic properties, genetic engineering, current regulatory framework and future prospects of bioplastic. Further the applications of bioplastics in industrial sector as well as opportunities and role of bio plastic in green economy are also discussed.

Keywords: algae, bioplastics, biodegradable, sustainability, green economy

1. Introduction

Plastics are synthetic carbon-based polymers that offer a variety of applications in everyday life. Petroleum-based plastics are widely used due to their desirable characteristics, which include low-cost, transparency, light-weight, strong heat resistance with a good weight-to-strength ratio. It can be easily shaped into different forms to produce a variety of materials. Thus, it has a wide range of applications in the expanding industrial sector. Overuse of these fossil-based, non-biodegradable polymers has shown serious impact on the environment, resulting in pollution, global warming and fossil fuel depletion. They have a poor recyclable potential and moreover, produce toxins in the recycling process. Furthermore, the plastic recycling process is quite challenging, as different plastics require different recycling techniques. Only around 10% of the total plastic manufactured every year is recycled, with the rest dumped into the water bodies and landfills [1, 2]. The indigenous microorganisms do not have inherent potential to degrade these plastic wastes. As a result, both terrestrial and aquatic ecosystems are severely harmed. Bioplastic made from renewable sources with similar qualities of fossil-based polymers is a viable alternative to overcome these major challenges. They are biodegradable and environment friendly, which can lead to a more sustainable and circular economy. In addition, bioplastics can be entirely decomposed by soil microorganisms without

IntechOpen

producing any harmful by-products [3]. A range of polymers utilised in bioplastic synthesis can be derived from key metabolites such as lipids, proteins, and carbohydrates. There are microbes that can utilise these polymers as a source of carbon and energy for their metabolism, and there are species that can produce exoenzymes to degrade them [4]. Given the fact that bioplastics can be made from a variety of renewable sources such as higher plants, bacteria, starch, and cellulose-based materials, algal plastics remain a high-demand research due to multiple advantages of algae as a feedstock.

Algae and cyanobacteria, also known as prokaryotic microalgae, are photoautotrophic organisms that grow at a faster ratewith high biomass. They are popular in the field of bioplastics because of their limited nutritional requirements, harvest regardless of the season and ability to thrive in non-arable environments, including waste waters [5]. Algaecan assimilate carbon dioxide into various organic compounds, which can then be transformed into useful biopolymers, resulting in reduced CO2 emissions, and ultimately leading to a safer environment [6, 7]. Algal biomass, in addition to algal metabolites, can be utilised directly in plastic production, where it can be blended with petroleum-based plastics to improve its mechanical qualities [8]. Algal biomass can also help to speed up the decomposition process of plastic, owing to its high nitrogen concentration, which improves microbial adhesion and promotes the biofilm formation [9]. Each of these beneficial attributes indicates algae as a potential future feedstock for bioplastic production. This chapter summarises all aspects of algae bioplastics and their function in ecological sustainability. This aims to assist and realise the relevance of bioplastic research, as well as the obstacles it faces and the necessity to overcome them in the future.

2. Properties of bioplastics

Bioplastics, as the name implies, are bio-based, biodegradableand compostable polymers containing mechanical and barrier attributes comparable to petroleum-based plastics. There are a variety of starch-based, cellulose-based, and protein-based bioplastics in the market today, but most of them are derived from food crops, which compete with human consumption. Poly Lactic Acid (PLA) and Polyhydroxyalkanoates (PHAs) are two typical algal biopolymers that have advantages over plant-based bioplastics as algae are easy to cultivate, non-competitive with human food, and can be harvested throughout the year [10]. Biodegradability of these polymers isdetermined by their structures, and strong mechanical qualities make them suitable for industrial uses. Bioplastics also have the potential to be customised in terms of properties, making them far preferable to conventional plastics. **Table 1** shows some of the advantages and disadvantages of bioplastics. Thermal stability, tensile strength, viscosity, elasticity, oxygen permeability, and water resistance are some of the significant characteristics of bioplastics.

Advantages of bioplastics	Disadvantages of bioplastics	Reference
Energy efficient	Expensive	[10–13]
Flexible to be modified	Brittle	
Do not generate toxic by products	Low melt strength	
Biodegradable	Weak barrier properties	

Table 1.
Advantages and disadvantages of bioplastics.

Numerous studies have proven that adding additives such as plasticizers and fillers to bioplastics strengthen both their structural and mechanical features [14]. When used as fillers, algal biomass itself has the potential to improve biodegradability. Kalita et al. [9] studied the biodegradation abilities of PLA material with algae biomass as a filler, and found that it increased the bioplastic's biodegradability. Hydrolysis of ester groups into hydroxyl or carboxyl groups occurs during PLA degradation. The algal biomass and PLA composites were extruded into a film and subjected to abiotic and composting degradation conditions. Water hyacinth compost set up was constructed for compost degradation experiments, and the films were cut into pieces and placed in 1 M NaOH for abiotic degradation. The sudden drop in molecular weight under abiotic stress conditions symbolises the molecule's degradation, which was seen in the presence of algal bio fillers. In the test setup with algal fillers, the degradation in the compost conditions was also noticeable, with days required for total biodegradation decreased from 95 ± 7 to 60 ± 2 days. These experiments clearly demonstrate the effectiveness of algae in improving degradation due to its high nitrogen concentration, which attracts microbes [9]. When no additives are used, Starch-based bioplastics have strong biodegradability, but poor mechanical stability compared to traditional plastics [11]. Methods like Coating, blending, as well as physical and chemical alterations can all be used to enhance their properties making them a complete sustainable alternative. Coating is the process of applying a topcoat of materials such as polycaprolactone and polyethylene oxide to assist improved barrier properties, tensile strength and elasticity. Nanomaterials, cellulose, thermoplastic starch, polycaprolactone are common compounds used in blending. When polymers are blended with nanomaterials, the polymer becomes confined between the nanoparticles, resulting in better barrier properties. Cellulose and thermoplastic starch combines well with other biopolymers, lowering water permeability while increasing mechanical qualities such as tensile strength. Polycaprolactones decrease polymer brittleness while simultaneouslyimproving heat stability [15].

Thummala et al. [12] examined the effects of glycerol and sorbitol as plasticizers on protein-based polymers. The findings show that sorbitol enhanced tensile strength whereas glycerol and a combination of the two showed intermediate tensile strength, indicating that bioplastics can be altered to meet specific requirements [12]. Studies on the effect of mould temperature on the viscosity of algal biopolymers demonstrate that increase in mould temperature improves viscosity and water resistance [13, 16, 17]. Plasticizers such as glycerol, water, and latex can enhance antibacterial properties of bioplastics, overcoming bioplastic's limitations in the medical and food packaging industries [18]. Antimicrobial additives like Nisin and cinnamaldehyde increase mechanical qualities and do not interfere with biodegradability, which was formerly a major concern [19, 20]. Although many of the approaches for improving characteristics are performed on plant-based plastics, they can all be applied to algal biopolymers, thus expanding the scope of algal research.

3. Algae used in bioplastic production

Algae are diverse group of photosynthetic organisms that range in size from single-celled microalgae to multicellular macroalgae which play a key role in ensuring a balanced ecosystem. They produce high metabolite content which can be processed into a number of value-added products, offering them to obtain wide range of market opportunities [21]. Bioplastics are one such products derived from algal metabolites, with the most notable being polylactic acid (PLA), polyhydroxyalkanoates (PHAs). These metabolites are naturally formed by algal cells, however the addition of particular chemicals, changes in culture conditions, can help to enhance the metabolite production.

One of the most commonly studied PHAs in bioplastic research is polyhydroxy butyrate (PHB). Sodium acetate in the culture mediumcan enhance PHB accumulation in cells without interfering with cell multiplication [22]. Kavitha et al. [23] optimised culture conditions for PHB production using wastewater cultured *Botryococcusbraunii*. According to the data, the maximum production of PHB (247 ± 0.42 mg/L) was obtained at a 60% concentration of sewage water as culture medium at 40°C and pH 7.5 [23]. Mathiot et al. [24] performed study on selection of microalgal strains for the production of starch-based bioplastics. They selected ten microalgal strains, including *Chlamydomonas*, different species of *Chlorella* and *Scenedesmus*. The algal strains were grown in a sulphur-depleted TAP medium with an 18-hour light: 6-hour dark photoperiod with a light intensity of 125 $\mu mol.m^{-1}.s^{-1}$ at 25°C and pH between 7.2–7.4 They conclude that *Chlamydomonasreinhardtii* is the best strain for bioplastic production among the 10 strains tested, with a starch to biomass ratio of 49 percent in a sulphur-depleted medium and excellent plasticization with 30 percent glycerol at 120°C [24]. In comparison to synthetic media, a study on optimising culture conditions on *Chlorella Salina* to discover the best photoperiod, CO2 concentration, and nutrient limitations suggests wastewater under controlled conditions as a feasible medium for optimum starch and carbohydrate production. The findings shows that a 12 h: 12 h light–dark cycle, 5% v/v CO2 concentration, and a combination of nitrogen and phosphorous limits is the ideal culture condition with highest starch and carbohydrate content. However, as compared to culture in synthetic medium, wastewater aerated with 5% CO2 is considered more sustainable, with double metabolite accumulation [25]. According to Das et al. [26] the leftover algal biomass after extracting lipids can also be used in bioplastic production and the result shows 27% PHB content. *Chlorella pyrenoidosa* was the strain under study in Fogg's medium with 80 Lux light intensity and UV spectroscopy analysis confirms PHB accumulation [26]. Since microalgae are low-cost substrate, Khomlaem et al. [27] used defatted *Chlorella* biomass and used as a substrate for three bacterial strains to accumulate PHAs. *Cupriavidusnecator*, a bacterial strain employing 75.4 percent defatted chlorella biomass, demonstrated the highest PHA accumulation (7.51 + 0.20 g/L) among the studied strains [27]. Another analysis revealed that adding glycerol as a plasticiser with defatted *Chlorella* biomass (DCB) to make Chitosan-based biodegradable films improved mechanical characteristics. Higher DCB concentrations resulted in increased tensile strength, reduced water vapour permeability, and reduced transparency. According to FTIR and SEM analysis, the increased attributes were because of the uniform distribution of DCB, which establishes strong hydrogen bonds throughout the matrix [28]. When employed as a carbon source, DCB can also be used to accelerate PHA and carotenoids production [29]. *Scenedesmussps*., that has not been studied extensively in the field of bioplastics, could also be a source of PHA in modified nutritional conditions. Since it has a rigid cell wall, it can withstand high temperatures, pH, copper concentrations, and a certain amount of salt [30, 31]. The majority of the studies described are recent findings, and the most of them focused on *Chlorella sps*. Because of its rapid growth rate and ease of cultivation.

4. Cyanobacteria used in bioplastic production

Cyanobacteria, commonly knownas blue green algae, are gram-negative prokaryotes with a wide range of species. In comparison to microalgae, they are well recognised for their PHA accumulation under stress conditions and are extensively studied for bioplastic production. In combination with microalgae and heterotrophic bacteria, cyanobacteria can be grown in municipal and industrial waste waters. They can grow rapidly in wastewater systems and serve a variety of different functions such as heavy metal removal, reduction of Chemical Oxygen Demand (COD) and Biological Oxygen Demand (BOD). For commercial production, a two-stage cyanobacterium cultivation can also be employed, with a photoautotrophic phase followed by a heterotrophic stage with a sole source of carbon [32]. In addition, filamentous cyanobacteria such as *Pseudanabaenasps.* and *Aphanocapsasps.* can survive in nutrient deprived conditions and can also outcompete rapidly growing green algae like *Chlorella* [33]. *Cyanobacteria* can be blended with other bioplastics, their smaller size facilitates uniform dispersion, which helps to form cross linkages and increase mechanical qualities. *Spirulina* is one of these species, which has been widely used in the food industry for many years due to its high protein content and has recently been exploited in the bioplastic research with promising results [34]. The addition of Polyvinyl alcohol as a compatibilizer and glycerol as a filler to a spirulina-based bioplastic increased the tensile strength by 1.89 kgf/cm2 and the elongation by 4.17 percent over commercial plastic [35]. When cultivated in modified Zarrouk medium, where the manufacturing cost was quite

Strain	Culture conditions	Biopolymer	Yield (% dry cell weight)	Reference
Nostocmuscorum	Acetate in medium +Dark incubation	PHB	43%	[42]
Synechococcus sp. MA19	Autotrophy, Phosphate deprivation	PHB	55%	[43]
Synechocystis sp. PCC6803	glucose containing BG11(Pre-grown) medium+Acetate+ Phosphhate deprivation	PHB	29%	[44]
Synechococcussubsalsus	Nitrogen deprivation	PHAs	16%	[45]
Spirulina sp. LEB-18	Nitrogen deprivation	PHAs	12%	[45]
Spirulina subsalsa	Increased salinity+ Nitrogen deprivation	PHA	7.45%	[46]
Calothrixscytonemicola	Photoautotrophy in nitrogen limitation	PHB	25.4 ± 3.5%	[47]
Aulosirafertilissima	Acetate and citrate supplemented medium	PHB	66%	[48]
Aulosirafertilissima	Acetate supplementation+ phosphate deprivation	PHB	77%	[48]
Anabaena cylindrica	Acetate supplemented BG11 medium	PHB	2%	[49]
Spirulina maxima	Acetate supplemented mixotrophic conditions	PHB	3%	[50]
Microalgal consortium	Waste water	PHA	43%	[51]

Table 2.
Microalgae and cyanobacteria used in bioplastic production.

high, spirulina-based plastic had higher biodegradable properties with 6.2 percent PHB content [36]. In contrast, a research of thermoplastic blends of bioplastics with *spirulina* grown in wastewater found that spirulina outperforms *Chlorella* as a blend. As a result, it is considered as more suitable for commercial purposes [37]. *Spiriulina platensis* showed increased thermal stability without influencing water vapour transmission rate when used as a filler in wheat-gluten based bioplastic [38]. Arias et al. [39] studied the effect of nitrogen and phosphorous starvation in two photoperiods, one with full light and the other with 12 h alternate light and dark, on a mixed culture dominated by cyanobacteria. According to their findings, Nitrogen limitation under alternate light illumination yielded the maximum carbohydrate concentration of 838 mg/L [39]. *Synechocystissps.* Showed increased PHB accumulation by up to 38 percent when cultivated in the presence of fructose and acetate in phosphorus deficient and gas-exchange limited conditions, which was eight times higher than accumulation under autotrophic conditions [40]. However, a study on *Arthrospira platensis* found that limiting nitrogen and phosphorus in the medium at the same time resulted in lower PHB and Phycocyanin synthesis. This research was carried out in both autotrophic and mixotrophic conditions, in both normal and nutrient-limited environments. Although autotrophic conditions with more CO2 resulted in increased PHB production of 33%, there was no significant increase in other conditions. The results were justified by stating that the effects of nutrient limitation conditions may differ from species to species, and that limiting both nitrogen and phosphorous at the same time would not be a good option in the case of *Arthrospira* [41]. The very first research on PHA synthesis in *Synechococcus elongates* was successful and exhibited maximum production of 17.15 percent of PHA under nitrogen starvation and phototrophic conditions with 1 percent sucrose as external carbon source. However, its yield was lower than bacterial systems and can be enhanced by applying alternative nutrient deprivation conditions and by using genetic tools (**Table 2**) [52].

5. Biopolymers derived from algal metabolites

Algal biomass, in addition to algal metabolites, can be directly moulded into bioplastic beads or sheets. When actively growing cultures are centrifuged, TAG accumulates and settles in the pellet. The pellet can be combined with additives like glycerol before being formed into the appropriate bioplastic shape [53].

5.1 Polyhydroxyalkanoate (PHA)

PHAs are biopolyesters produced as intracellular inclusions by a variety of microorganisms, especially in the presence of abundant carbon and limited essential nutrients. They accumulate in the cell as it enters the stationary phase and can account for up to 80% of the cell's weight. These inclusions are protein and lipid-based membrane bound inclusions. They serve as energy reserves for the cells, allowing them to endure oxidative stress, UV irradiation, temperature shock, and osmotic imbalance. PHAs are made of polyhydroxyalkanoic acid monomer units in which the ester bond is between the carboxyl group of one monomeric and hydroxyl group of the next monomeric unit. The R group in each monomeric molecule forms an alkyl side chain. Monomers can differ depending on the organism's substrate, resulting in the formation of various polymers and copolymers [5, 54]. They are UV resistant, insoluble, and have low oxygen permeability. Melting temperatures range from 40 to 180°C, with a glass transition temperature of −50 to 40°C. The temperature ranges stated here differ depending on the R-group [55]. Cyanobacteria

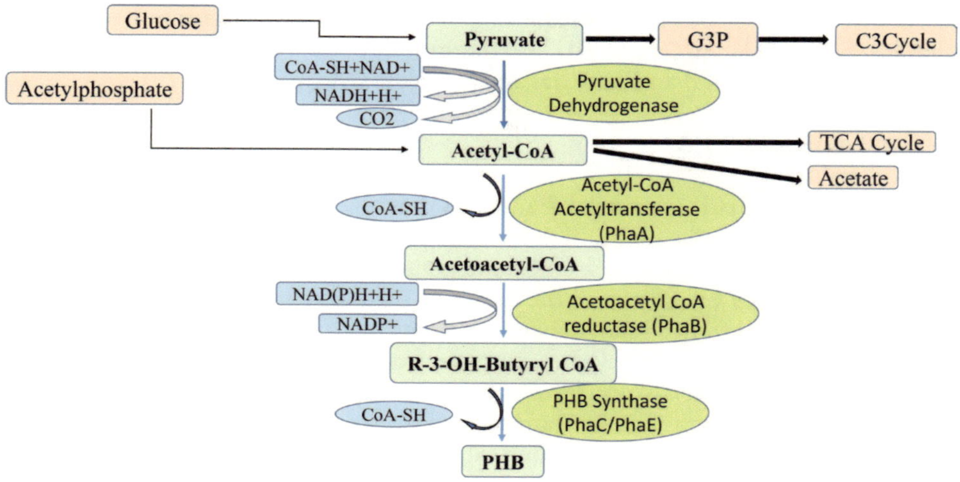

Figure 1.
Metabolic pathway of PHB synthesis showing acetyl-CoA as a branch point.

are known as best producers of PHA and are being constantly studied for ways to boost synthesis in nutrient-restricted conditions or through genetic engineering experiments. Polyhydroxy butyrate (PHB) is the well-known and widely marketed PHA which has qualities similar to petroleum-based polymers [56]. It's made up of repeating units of 3 carbon atoms and a methyl group [55]. **Figure 1** depicts the PHB synthesis metabolic route with acetyl-CoA as a branch point. PHB production can be improved by altering the enzymes in the pathway, either by boosting PHB synthesis enzymes or by deletions of other Acetyl-CoA consuming enzymes to increase the substrate for PHB synthesis.

Acetyl CoA is a branch point in microalgae's central metabolism. In the cells, glucose is transformed to pyruvate, which can then be converted to Glyceraldehyde 3-Phosphate (G3P) and enter the Calvin Cycle (C3 cycle) to produce carbohydrates. The enzyme pyruvate dehydrogenase transforms pyruvate to acetyl CoA, which is the primary substrate for PHB synthesis. Conversion of Acetyl phosphate is another source of Acetyl CoA synthesis. Acetyl CoA synthesised can be used by the cells in PHB synthesis, Tri carboxylic acid (TCA) cycle, or converted to acetate. A set of enzymes, including PhaA, PhaB, and PhaC/E, convert acetyl CoA to PHB in the PHB production pathway [57, 58].

5.2 Polyurethane (PU)

Polyurethane is commercially available biopolymer widely used in adhesives, coatings, elastomers, and foams [59]. They can be made from oils, in which algal oils, such as triglycerides are sustainable sources. The composition of fatty acids varies from species to species. Pawar et al. employed *Chlorella* to produce polyols through oxidation. The oil had a 10% saturates composition and a 2% unknown fatty acid content. Algal oil epoxidation yielded good results, with conversion rates comparable to other vegetable oils. The epoxide ring opening of epoxidized algal oil with ethylene glycol and lactic acid was successfully achieved for the manufacture of rigid polyurethane foams. The synthesis resulted in the manufacture of polyurethane with characteristics similar to those of commercially available polyols [60]. Patil et al. [61] prepared nanocomposite coatings using Polyol from algal oil and ricinoleic acid by combining eggshell based Silver doped nanoparticles in a Polyurethane matrix. The qualities of PU nanocomposite coatings were compared to PU without nanoparticles and according to the findings, PU coatings had good physico-mechanical properties [61]. Furthermore, the polyurethane coatings made from algal oils exhibited antibacterial and anticorrosive qualities [62].

5.3 Polylactic acid (PLA)

PLA is usually made from algal feedstocks that is fermented to produce lactic acid and then polymerised [63]. PLA is one of the most efficient plastics since it uses a less amount of feedstock (sugar) to generate a biodegradable plastic. In comparison to other biopolymers, the rate of CO2 release is also lesser. It's one among the polymers whose stereochemical structure can be easily changed to get a higher molecular weight. Other characteristics, such as amorphousness and semi-crystalline nature, can also be altered by varying isomers [64]. Since it is an FDA-approved biopolymer, it is commonly utilised in food packaging. Because the monomer is chiral and exists in two optical isomeric states, various PLA structures can be formed. Poly (L-Lactide) (PLLA), Poly (D-Lactide) (PDLA), and Poly (D, L-Lactide) are the three. Packaging made of poly (D, L lactide) with 90% L-Lactide is commonly utilised. Increased D-Lactide in the composition results in a polymer with a high crystalline structure with good mechanical and barrier properties, but it is not economically successful due to its high cost [65].

5.4 Cellulose acetate

Cellulose is the most prevalent natural polymer on the planet, making it a limitless source of raw material for creating environmentally beneficial products without interrupting the food chain. It is made up of glucose monomer units linked together by a β-1,4 glycosidic linkage that compacts them and forms strong inter-chain hydrogen bonds. The structure's alternating side chains contribute to its high crystalline nature, which causes brittleness, poor flexibility, and weak tensile strength. But, in combination with appropriate plasticizers, cellulose derivatives such as ethers and esters act as bioplastics with a range of applications. One such polymer is cellulose acetate, which is used in the production of items such as spectacle frames, combs, cigarette filters and disposable jewellery. Due to its dimensional stability, rigidity, and printability, cellulose acetate was also recognised for its usage in food packaging, where it was used to wrap baked products. However, the packaging industry no longer prefers cellulose acetate due to its weak moisture and gas barrier attributes, as well as the fact that it will be hydrolysed to produce acetic acid. The production of cellulose acetate on a large scale is normally done under regulated conditions, particularly the temperature, which impacts the degree of polymerisation (DP). The product quality will be affected if the DP is too low [66, 67]. Although there are numerous studies on the manufacturing of cellulose acetate from plant sources, research on the production of cellulose acetate from algal cell walls are still under progress. However, because the cell walls are not entirely formed of cellulose, the yield of cellulose-based polymers will be low. But the large-scale production in collaboration with other biopolymers using the remaining biomass will be a viable option to use the complete biomass.

6. Genetic engineering for improved metabolite production

Genetic engineering is a sophisticated method for gene manipulation that has been utilised in a number of research studies. In terms of bioplastic manufacture, there have been various investigations on plant gene manipulations, which are referred to as first- and second-generation bioplastics. However, research has now shifted to third-generation bioplastics generally known as algal bioplastics. Genetic studies on algae and cyanobacteria are easier due to their lower complexity compared to plants, and they have a high potential for producing bioplastic. *Synechosystis sp*. PCC6803 is the first photosynthetic organism to be completely sequenced, which is one of the reasons it is being explored in gene manipulation for increased PHA production [54]. PHB accumulation is known to be increased in nitrogen-deficient conditions, and the Sigma factor *sigE*is known for its potential to activate numerous carbohydrate metabolic pathways, including the PHB synthesis pathway. Under nitrogen-deficient conditions, overexpression of *sigE* in *Synechocystis sp*. PCC 6803 resulted in increased PHB synthesis. Importantly the molecular weight and monomer units of the produced PHB are identical to those of wild-type PHB [68]. Transformation experiments on *Synechocystis sp*. PCC 6803 containing *pha* genes were also effective, with the resultant cells accumulating 12-fold higher PHB under nitrogen stress than the wild type strain [69]. Acetyl-CoA is a branch point in algal cell's core metabolism, and it can be transformed into a number of substances based on the cell's need via various enzymes, including PHB synthesis. Phosphotransacetylase and acetyl-CoA hydrolase are enzymes that convert acetyl-CoA to acetate and are encoded by the *pta* and *ach* genes, respectively. This reduces the amount of substrate available for PHB synthesis. Phosphoketolase, on the other hand, is an enzyme produced by the *xfpk* gene that can increase acetyl-CoA levels in the cell. Carpine et al. [70] employed a different approach to boost PHB production, instead of overexpressing the enzyme in the synthesis pathway, they aimed to increase the substrate concentration for PHB synthesis. This experiment was designed by engineering *Synechocystis sp*. PCC6803. Seven mutants were constructed with three different gene alterations, including deletions of the pta and ach genes and overexpression of the xfpk gene. They were effective, and their findings demonstrate that the mutant with all three modifications accumulated the most PHB (232 mg/L) [70]. Orthwein et al. [71] discovered a novel protein, pirC (PII-interacting regulator of carbon metabolism), and investigated its function for PHB production. The PirC deficient mutant strain of *Synechocystis* found to have higher phosphoglycerate mutase activity, leading to increased PHB production. The strain was modified even more by transferring PHA metabolism genes (phaA and phaB) from known PHB producing bacteria, *Cupriavidus necator*, which also showed good results and was termed PPT1. The strain produced 63 percent PHB in nitrogen and phosphorus limitation. PHB level increased to 81 percent in the presence of acetate under the same culture conditions, making it the highest PHB content to be reported in any known cyanobacterium [57]. Although cyanobacteria are widely used in genetic engineering studies, Hempel et al. [72] incorporated a *Ralstoniaeutroph*a bacterial PHB synthesis pathway into the diatom *Phaeodactylumtricornutum*, demonstrating microalgae as a workable model for PHB production. These bacterial enzymes were sufficient to synthesise PHB in the cells, accounting for up to 10% of the dry weight of the algae. This research was one of the first to utilise genetic engineering to produce PHB in microalgae, and it cleared the path for further research [72]. CRISPR/Cas systems are gene editing tools that can produce a variety of mutant, knock-out, and knock-in strains with desired characteristics. This approach has yet to be thoroughly investigated in algal systems, particularly in bioplastic research [73].

7. Applications of bioplastics

Plastics have a range of applications, and bioplastics have the potential to replace conventional plastics in all of them. Bioplastics can be moulded in a variety of ways, from fibre to thin film, and can be designed in any size, shape, or dimension. PHB is most likely to be widely used in food packaging, whether it is for fresh or long-term storage. Green house films, protection nets, and grow sacks are examples of bioplastics used in agriculture to maintain appropriate conditions and protect the crop from physical and biological risks. Unlike synthetic polyethylene, these grow bags do not cause deformity, making them root friendly. Since PHA is a biomaterial, it can interact with biological system and elicit a favourable response from the host. As a result, it has applications in medicine, including in the engineering of biological tissues such as bone, cartilage, and skin. It can be utilised to regenerate dental tissue [2], employed for fraction fixationas well as surgical sutures. The medicament can be loaded into PHB-based wound dressings, and the fibrous nature of the material facilitates the drug's release into the wound. The property of cancer cells adhering to PHB sheets has been documented, and contact angle techniques can be used to identify it. In comparison to biopsy, which is often used for medical examinations, this approach is painless [74]. PHB with a high molecular weight can be employed as a drug carrier [23, 75]. Because of its non-adhesive qualities, PHB can be utilised as an effective antibacterial agent in aquaculture, preventing pathogens from forming biofilms and thereby inhibiting infection. PHB is an antifouling compound that can be blended with other metals and applied to the hull to prevent undesired marine creatures from settling [58]. PLA blends are used in a variety of applications, including computer and mobile phone casings, medical implants, and various packaging materials such as cups, tins, and bottles [64]. Algal plastics can also be used to make plastic beads, which have applications in fishing, ornament crafting, and shooting sports. However, it is not much explored due to the high expense of extraction and purification of biopolymers. But research on *Chlamydomonas reinhardtii* showed that triacylglycerol can be directly moulded into 7 mm beads without the use of extraction or purification [53].

8. Challenges in the field of bioplastics

Microalgae are well-known for their ease of cultivation, capacity to grow in waste waters, and ability to survive in adverse environments, but producing a marketable product from this biomass has numerous hurdles at each step along the way, from cultivation to market release. Not all species are capable of adapting to a wide range of cultural settings. As a result, the strain chosen for research may not be adapted to the designed conditions, making the cultivation phase challenging. Following cultivation, the harvesting process, which is highly costly, presents the next hurdle. Major research focuses on ways to boost metabolite synthesis, but there is also an urgent need to identify cost-effective harvesting methods. In addition, the biomass and metabolites produced will be insufficient for industrial production. This difficulty can be solved by inducing heterotrophic conditions with an external carbon source, however there is a high risk of contamination [76]. Genetic engineering is well known for its excellent outcomes in terms of increased metabolite production but, it also offers several challenges, including the need for a genome sequence, the difficulty in gene alterations, and the maintenance and genetic stability of transgenic strains [77]. Moreover, as transgenic cyanobacteria can pose an ecological damage, they cannot be grown in open systems [54]. Mutagens can be used to cause random mutations, which can be a suitable alternative to genetic engineering. But this necessitates extensive screening in order to find the mutant with the desired properties [73].

Cyanobacteria-dominated mixed cultures are also known for their high PHA production, although maintaining their dominance without contamination is difficult prior to purification. Because Cyanobacteria can tolerate high nitrogen concentrations, maintaining high N: P ratios is considered to be a viable solution. But, mixed cultivations have only been used in laboratory and pilot scale manufacturing, and scaling up production will probably take longer [33]. When compared to currently available polymers, pure algae bioplastics have lower mechanical strengths, which limit their uses. However, employing sustainable biomass as additives such as compatibilizers, fillers, and plasticizers is a possible approach for addressing this problem [78]. Biodegradability is the significant property of bioplastics due to which, it is on high demand. But this requires a set of conditions that may not be present in landfills, where they are usually disposed. As a result, bioplastics that degrade under normal conditions will need to be tailored in the future. Furthermore, if not properly disposed of, bioplastics might emit a small amount of greenhouse gases. These, gases on the other hand, can be collected and used for other purposes, such as biogas production [79]. Future algae research should focus on finding alternatives to all of the aforementioned issues, with the objective of enhancing large-scale production.

9. Role of algal bioplastics in green economy

Unlike conventional plastics, where the whole process remains linear and continuously emit harmful gases including CO2, Algal bioplastics play a significant role in building a future green economy by using emitted CO2 as the carbon source for their survival making it a circular process. Only around 1% of the world's plastics are biodegradable, while the rest are fossil-based, posing a threat to significant flora and fauna in both terrestrial and aquatic habitats. If this condition persists, it will harm the species on the globe and may even result in the extinction of species, reducing biodiversity. This shows the worldwide impact of plastic use, which can be mitigated by the use of sustainable algae bioplastics.

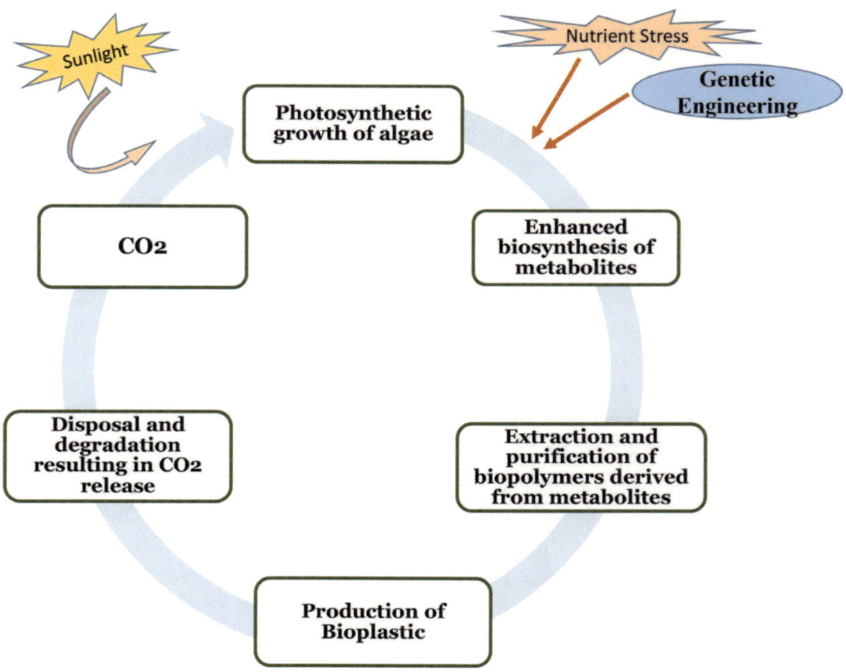

Figure 2.
Role of algal bioplastics leading to a circular and green bioeconomy.

Figure 2 is a representation of an eco-friendly cycle as a result of using algal bioplastics. Algae use sunlight and CO2 as raw materials for photosynthesis, producing and accumulating a variety of metabolites that can be improved by manipulating nutritional conditions, using gene editing, or inducing mutations. These metabolites can be transformed into biopolymers, which can then be isolated and purified to make bioplastics. After disposal, these bioplastics degrade to produce CO2, which is then used by algae, and the cycle continues. Further, study and improvements onbioplastics and eventually replacing plastics entirelywill result in a green and safer planet.

10. Future prospects of bioplastic production

Despite the fact that this chapter summarises the negative effects of conventional plastics, the benefits of bioplastics, the research framework for bioplastic production and the role of algal bioplastics in the green economy, the majority of the studies are conducted at the laboratory or pilot scale. Only large-scale production, as well as other species characterisation studies will be able to meet the increasing demand. To begin with, algae are highly diverse and there are many species and classes of microalgae that are yet to be identified. The selection of species for an experiment is critical, and further research on evolutionary divergence and species categorisation should be done beforehand to explore the properties and efficiency of each species. Increased metabolite accumulation in response to abiotic stress is well-known, and nitrogen and phosphorus deficiency are particularly well-studied. Although the stress-induced increase in metabolite accumulation is true, it has an adverse influence on the cells and decreases the biomass rate. To overcome this challenge, a two-stage cultivation approach can be used. In this method, algal cells are first grown in optimal conditions before being stressed to accumulate metabolites [32]. This results in higher biomass and metabolite yields at the same time. Adopting these strategies in large-scale production, on the other hand, is less prevalent. Genetic engineering is another cutting-edge method for enhancing metabolite production. To target a transcription factor or a gene involved in metabolic pathway, a variety of genome editing approaches can be applied. This is a highly successful strategy that has yielded positive results in laboratory trials [80, 81]. Nevertheless, since algae are complex eukaryotic organisms with few genomes sequenced, most researchers are limited to studying only those organisms that have had their genomes sequenced. To explore algal genetics, independent research on algal genome sequencing is required. These genetically engineered organisms can also be grown in nutrient-deficient conditions to enhance yields even further. However, algal genetic engineering experiments are limited to laboratory study, and these genetically engineered organisms are not permitted to be cultivated for industrial metabolite production. This is in regards to the stability and maintenance of the strain, as well as the possibility of lateral gene transfer. As a result, techniques for maintaining stability while also safely disposing of remaining genetic material are required. Mixotrophic cultures are now being studied for metabolite production, with promising results. This can also be utilised in large-scale cultivations to overcome the challenges of maintaining pure cultures. This method is particularly useful in waste water treatment, as these mixotrophic cultures can develop by utilising the extra nutrients in the waste waters while also result in waste water treatment. Finally, using the above mentioned strategies for improving metabolite production can either be employed separately or coupled with different combinations for large scale production with profitable yields.

11. Conclusion

Excessive usage of plastics results in pollution, causing harm to the earth and its existing species. Despite the numerous benefits of algae bioplastics, research in this area still need to be progressed. There is also a critical need to take advantage of modern genetic technologies to boost the metabolite synthesis for bioplastic production. Because there have been proven results of employing various algal and cyanobacterial strains as additives that demonstrate increased mechanical qualities of bioplastics equivalent to adding other synthetic components, using solely these renewable sources helps to develop highly compostable bioplastic.

Acknowledgements

Authors would like to thank *Science and Engineering Research Board, Govt. of India for SERB-SRG research grant. We would also like to thank Department of Microbiology, School of Life sciences, Central University of Tamil Nadu for supporting this work.*

Conflict of interest

Authors declared that there is no conflict of interest with respect to either authorship, affiliation or in any part of the writing of this chapter.

Author details

Arathi Sreenikethanam and Amit Bajhaiya*
Algal Biotechnology Lab, Department of Microbiology, School of Life Sciences, Central University of Tamil Nadu, Thiruvarur, India

*Address all correspondence to: amitkumar@cutn.ac.in

References

[1] Thiruchelvi R, Das A, Sikdar E. Bioplastics as better alternative to petro plastic. Materials Today: Proceedings. 2021 Jan 1;**37**:1634-1639

[2] Sidek IS, Draman SF, Abdullah SR, Anuar N. Current development on bioplastics and its future prospects: an introductory review. INWASCON Technology Magazine. 2019;**1**:03-08

[3] Kadar NA, Rahim NS, Yusof R, Nasir NA, Hamid HA. A REVIEW ON POTENTIAL OF ALGAE IN PRODUCING BIODEGRADABLE PLASTIC. International Journal of Engineering Advanced Research. 2021 Mar 1;**3**(1):13-26

[4] Trivedi P, Hasan A, Akhtar S, Siddiqui MH, Sayeed U, Khan MK. Role of microbes in degradation of synthetic plastics and manufacture of bioplastics. Journal of Chemical and Pharmaceutical Research. 2016;**8**(3):211-216

[5] Costa SS, Miranda AL, de Morais MG, Costa JA, Druzian JI. Microalgae as source of polyhydroxyalkanoates (PHAs)—A review. International journal of biological macromolecules. 2019 Jun 15;**131**:536-547

[6] Zabochnicka-Świątek M. Algae–feedstock of the future. ArchivumCombustionis. 2010;**30**(3): 225-236

[7] Rajendran N, Puppala S, Sneha Raj M, Ruth Angeeleena B, Rajam C. Seaweeds can be a new source for bioplastics. Journal of Pharmacy Research. 2012 Mar;**5**(3):1476-1479

[8] Rahman A, Miller CD. Microalgae as a source of bioplastics. InAlgal Green Chemistry 2017 Jan 1 (pp. 121-138). Elsevier.

[9] Kalita NK, Damare NA, Hazarika D, Bhagabati P, Kalamdhad A, Katiyar V. Biodegradation and characterization study of compostable PLA bioplastic containing algae biomass as potential degradation accelerator. Environmental Challenges. 2021 Apr 1;**3**:100067

[10] Machmud MN, Fahmi R, Abdullah R, Kokarkin C. Characteristics of red algae bioplastics/latex blends under tension. International Journal of Science and Engineering. 2013 Oct 10;**5**(2):81-88

[11] Nandiyanto AB, Fiandini M, Ragadhita R, Sukmafitri A, Salam H, Triawan F. MECHANICAL AND BIODEGRADATION PROPERTIES OF CORNSTARCH-BASED BIOPLASTIC MATERIAL. Materials Physics & Mechanics. 2020 Sep;**1**:44(3)

[12] Tummala P, Liu W, Drzal LT, Mohanty AK, Misra M. Influence of plasticizers on thermal and mechanical properties and morphology of soy-based bioplastics. Industrial & Engineering Chemistry Research. 2006 Oct 25;**45**(22):7491-7496

[13] Jiménez-Rosado M, Bouroudian E, Perez-Puyana V, Guerrero A, Romero A. Evaluation of different strengthening methods in the mechanical and functional properties of soy protein-based bioplastics. Journal of Cleaner Production. 2020 Jul 20;**262**:121517

[14] Arrieta MP, Peponi L, López D, López J, Kenny JM. An overview of nanoparticles role in the improvement of barrier properties of bioplastics for food packaging applications. Food Packaging. 2017 Jan;**1**:391-424

[15] Peelman N, Ragaert P, De Meulenaer B, Adons D, Peeters R, Cardon L, et al. Application of bioplastics for food packaging. Trends in Food Science & Technology. 2013 Aug 1;**32**(2):128-141

[16] González-Balderas RM, Felix M, Bengoechea C, Guerrero A, Ledesma MO. Influence of mold temperature on the properties of wastewater-grown microalgae-based plastics processed by injection molding. Algal Research. 2020 Oct 1;**51**:102055

[17] Jiménez-Rosado M, Rubio-Valle JF, Perez-Puyana V, Guerrero A, Romero A. Use of heat treatment for the development of protein-based bioplastics. Sustainable Chemistry and Pharmacy. 2020 Dec 1;**18**:100341

[18] Jones A, Mandal A, Sharma S. Protein-based bioplastics and their antibacterial potential. Journal of Applied Polymer Science. 2015 May;**10**:132(18)

[19] Balaguer MP, Villanova J, Cesar G, Gavara R, Hernandez-Munoz P. Compostable properties of antimicrobial bioplastics based on cinnamaldehyde cross-linked gliadins. Chemical Engineering Journal. 2015 Feb 15;**262**:447-455

[20] Perez-Puyana V, Felix M, Romero A, Guerrero A. Development of pea protein-based bioplastics with antimicrobial properties. Journal of the Science of Food and Agriculture. 2017 Jun;**97**(8):2671-2674

[21] Rahman KM. Food and high value products from microalgae: market opportunities and challenges. InMicroalgae biotechnology for food, health and high value products 2020 (pp. 3-27). Springer. Singapore.

[22] Abdo SM, Ali GH. Analysis of polyhydroxybutrate and bioplastic production from microalgae. Bulletin of the National Research Centre. 2019 Dec;**43**(1):1-4

[23] Kavitha G, Kurinjimalar C, Sivakumar K, Kaarthik M, Aravind R, Palani P, et al. Optimization of polyhydroxybutyrate production utilizing waste water as nutrient source by BotryococcusbrauniiKütz using response surface methodology. International journal of biological macromolecules. 2016 Dec 1;**93**:534-542

[24] Mathiot C, Ponge P, Gallard B, Sassi JF, Delrue F, Le Moigne N. Microalgae starch-based bioplastics: Screening of ten strains and plasticization of unfractionated microalgae by extrusion. Carbohydrate polymers. 2019 Mar 15;**208**:142-151

[25] Chong JF, Fadhullah W, Lim V, Lee CK. Two-stage cultivation of the marine microalga Chlorella salina for starch and carbohydrate production. Aquaculture International. 2019 Oct;**27**(5):1269-1288

[26] Das SK, Sathish A, Stanley J. Production of biofuel and bioplastic from chlorella pyrenoidosa. Materials Today: Proceedings. 2018 Jan 1;**5**(8):16774-16781

[27] Khomlaem C, Aloui H, Kim BS. Biosynthesis of polyhydroxyalkanoates from defatted Chlorella biomass as an inexpensive substrate. Applied Sciences. 2021 Jan;**11**(3):1094

[28] Deshmukh AR, Aloui H, Khomlaem C, Negi A, Yun JH, Kim HS, et al. Biodegradable films based on chitosan and defatted Chlorella biomass: Functional and physical characterization. Food Chemistry. 2021 Feb 1;**337**:127777

[29] Khomlaem C, Aloui H, Deshmukh AR, Yun JH, Kim HS, Napathorn SC, et al. Defatted Chlorella biomass as a renewable carbon source for polyhydroxyalkanoates and carotenoids co-production. Algal Research. 2020 Oct 1;**51**:102068

[30] HernodhSvantesson I. GrånäsJakobsson S. An analysis of creating plastic material based on the microalgae Scenedesmus.

[31] García G, Sosa-Hernández JE, Rodas-Zuluaga LI, Castillo-Zacarías C, Iqbal H, Parra-Saldívar R. Accumulation of PHA in the microalgae Scenedesmus sp. under nutrient-deficient conditions. Polymers. 2021 Jan;13(1):131.

[32] Monshupanee T, Nimdach P, Incharoensakdi A. Two-stage (photoautotrophy and heterotrophy) cultivation enables efficient production of bioplastic poly-3-hydroxybutyrate in auto-sedimenting cyanobacterium. Scientific reports. 2016 Nov 15;**6**(1):1-9

[33] Arias DM, García J, Uggetti E. Production of polymers by cyanobacteria grown in wastewater: Current status, challenges and future perspectives. New biotechnology. 2020 Mar 25;**55**:46-57

[34] OnenCinar S, Chong ZK, Kucuker MA, Wieczorek N, Cengiz U, Kuchta K. Bioplastic production from microalgae: A review. International journal of environmental research and public health. 2020 Jan;**17**(11):3842

[35] Dianursanti, Noviasari C, Windiani L, Gozan M. Effect of compatibilizer addition in Spirulina platensis based bioplastic production. InAIP Conference Proceedings 2019 Apr 9 (Vol. 2092, No. 1, p. 030012). AIP Publishing LLC.

[36] Maheswari NU, Ahilandeswari K. Production of bioplastic using Spirulina platensis and comparison with commercial plastic. Res. Environ. Life Sci. 2011;**4**(3):133-136

[37] Zeller MA, Hunt R, Jones A, Sharma S. Bioplastics and their thermoplastic blends from Spirulina and Chlorella microalgae. Journal of Applied Polymer Science. 2013 Dec 5;**130**(5): 3263-3275

[38] Ciapponi R, Turri S, Levi M. Mechanical reinforcement by microalgalbiofiller in novel thermoplastic biocompounds from plasticized gluten. Materials. 2019 Jan;**12**(9):1476

[39] Arias DM, Uggetti E, García-Galán MJ, García J. Production of polyhydroxybutyrates and carbohydrates in a mixed cyanobacterial culture: effect of nutrients limitation and photoperiods. New biotechnology. 2018 May 25;**42**:1-1

[40] Panda B, Mallick N. Enhanced poly-β-hydroxybutyrate accumulation in a unicellular cyanobacterium, Synechocystis sp. PCC 6803. Letters in applied microbiology. 2007 Feb;**44**(2):194-198

[41] Nematollahi MA, Laird DW, Hughes LJ, Raeisossadati M, Moheimani NR. Effect of organic carbon source and nutrient depletion on the simultaneous production of a high value bioplastic and a specialty pigment by Arthrospira platensis. Algal Research. 2020 May 1;**47**:101844

[42] Sharma L, Mallick N. Accumulation of poly-β-hydroxybutyrate in Nostocmuscorum: regulation by pH, light–dark cycles, N and P status and carbon sources. Bioresource technology. 2005 Jul 1;**96**(11):1304-1310

[43] Nishioka M, Nakai K, Miyake M, Asada Y, Taya M. Production of poly-β-hydroxybutyrate by thermophilic cyanobacterium, Synechococcus sp. MA19, under phosphate-limited conditions. Biotechnology letters. 2001 Jul;**23**(14):1095-1099

[44] Panda B, Jain P, Sharma L, Mallick N. Optimization of cultural and nutritional conditions for accumulation of poly-β-hydroxybutyrate in Synechocystis sp. PCC 6803. Bioresource technology. 2006 Jul 1;**97**(11):1296-1301

[45] Costa SS, Miranda AL, Andrade BB, de Jesus AD, Souza CO, de Morais MG, et al. Influence of nitrogen on growth, biomass composition, production, and

properties of polyhydroxyalkanoates (PHAs) by microalgae. International journal of biological macromolecules. 2018 Sep 1;**116**:552-562

[46] Shrivastav A, Mishra SK, Mishra S. Polyhydroxyalkanoate (PHA) synthesis by Spirulina subsalsa from Gujarat coast of India. International journal of biological macromolecules. 2010 Mar 1;**46**(2):255-260

[47] Kaewbai-Ngam A, Incharoensakdi A, Monshupanee T. Increased accumulation of polyhydroxybutyrate in divergent cyanobacteria under nutrient-deprived photoautotrophy: An efficient conversion of solar energy and carbon dioxide to polyhydroxybutyrate by Calothrixscytonemicola TISTR 8095. Bioresource technology. 2016 Jul 1;**212**:342-347

[48] Samantaray S, Mallick N. Production and characterization of poly-β-hydroxybutyrate (PHB) polymer from Aulosirafertilissima. Journal of Applied Phycology. 2012 Aug;**24**(4):803-814

[49] Lama L, Nicolaus B, Calandrelli V, Manca MC, Romano I, Gambacorta A. Effect of growth conditions on endo- and exopolymer biosynthesis in Anabaena cylindrica 10 C. Phytochemistry. 1996 Jun 1;**42**(3): 655-659

[50] De Philippis R, Sili C, Vincenzini M. Glycogen and poly-β-hydroxybutyrate synthesis in Spirulina maxima. Microbiology. 1992 Aug 1;**138**(8): 1623-1628

[51] Chakravarty P, Mhaisalkar V, Chakrabarti T. Study on poly-hydroxyalkanoate (PHA) production in pilot scale continuous mode wastewater treatment system. Bioresource technology. 2010 Apr 1;**101**(8):2896-2899

[52] Mendhulkar VD, Shetye LA. Synthesis of biodegradable polymer polyhydroxyalkanoate (PHA) in cyanobacteria Synechococcus elongates under mixotrophic nitrogen-and phosphate-mediated stress conditions. Industrial Biotechnology. 2017 Apr 1;**13**(2):85-93

[53] Kato N. Production of crude bioplastic-beads with microalgae: Proof-of-concept. Bioresource Technology Reports. 2019 Jun 1;**6**:81-84

[54] Koller M. "Bioplastics from microalgae"—Polyhydroxyalkanoate production by cyanobacteria. InHandbook of Microalgae-Based Processes and Products 2020 Jan 1 (pp. 597-645). Academic Press.

[55] Johnsson N. Steuer F. Bioplastic material from microalgae: Extraction of starch and PHA from microalgae to create a bioplastic material.

[56] Zhang C, Show PL, Ho SH. Progress and perspective on algal plastics–a critical review. Bioresource technology. 2019 Oct 1;**289**:121700

[57] Koch M, Bruckmoser J, Scholl J, Hauf W, Rieger B, Forchhammer K. Maximizing PHB content in Synechocystis sp. PCC 6803: a new metabolic engineering strategy based on the regulator PirC. Microbial Cell Factories. 2020 Dec;19(1):1-2.

[58] Kavitha G, Rengasamy R, Inbakandan D. Polyhydroxybutyrate production from marine source and its application. International journal of biological macromolecules. 2018 May 1;**111**:102-108

[59] Özçimen D, İnan B, Morkoç O, Efe A. A review on algal biopolymers. J. Chem. Eng. Res. Updates. 2017;**4**:7-14

[60] Pawar MS, Kadam AS, Dawane BS, Yemul OS. Synthesis and characterization of rigid polyurethane foams from algae oil using biobased

chain extenders. Polymer Bulletin. 2016 Mar 1;**73**(3):727-741

[61] Patil CK, Jirimali HD, Paradeshi JS, Chaudhari BL, Gite VV. Functional antimicrobial and anticorrosive polyurethane composite coatings from algae oil and silver doped egg shell hydroxyapatite for sustainable development. Progress in Organic Coatings. 2019 Mar 1;**128**:127-136

[62] Patil CK, Jirimali HD, Paradeshi JS, Chaudhari BL, Alagi PK, Hong SC, et al. Synthesis of biobased polyols using algae oil for multifunctional polyurethane coatings. Green Materials. 2018 Oct 9;**6**(4):165-177

[63] Kaparapu J. Polyhydroxyalkanoate (PHA) Production by Genetically Engineered Microalgae: A Review. J. New Biol. Rep. 2018 Aug 11;7:68-73

[64] Reddy RL, Reddy VS, Gupta GA. Study of bio-plastics as green and sustainable alternative to plastics. International Journal of Emerging Technology and Advanced Engineering. 2013 May;**3**(5):76-81

[65] Lackner M. Bioplastics. Kirk-Othmer Encyclopedia of Chemical Technology. 2000 Dec;**4**:1-41

[66] Stevens CV. Bio-based plastics: materials and applications. John Wiley & Sons; 2013 Oct 2

[67] Liu L. Bioplastics in food packaging: Innovative technologies for biodegradable packaging. San Jose State University Packaging Engineering. 2006 Feb;**13**:1348-1368

[68] Osanai T, Numata K, Oikawa A, Kuwahara A, Iijima H, Doi Y, et al. Increased bioplastic production with an RNA polymerase sigma factor SigE during nitrogen starvation in Synechocystis sp. PCC 6803. DNA research. 2013 Dec 1;**20**(6):525-535

[69] Hondo S, Takahashi M, Osanai T, Matsuda M, Hasunuma T, Tazuke A, et al. Genetic engineering and metabolite profiling for overproduction of polyhydroxybutyrate in cyanobacteria. Journal of bioscience and bioengineering. 2015 Nov 1;**120**(5):510-517

[70] Carpine R, Du W, Olivieri G, Pollio A, Hellingwerf KJ, Marzocchella A, et al. Genetic engineering of Synechocystis sp. PCC6803 for poly-β-hydroxybutyrate overproduction. Algal research. 2017 Jul 1;**25**:117-127

[71] Orthwein T, Scholl J, Spät P, Lucius S, Koch M, Macek B, Hagemann M, Forchhammer K. The novel PII-interactor PirC identifies phosphoglycerate mutase as key control point of carbon storage metabolism in cyanobacteria. Proceedings of the National Academy of Sciences. 2021;**118**(6):e2019988118

[72] Hempel F, Bozarth AS, Lindenkamp N, Klingl A, Zauner S, Linne U, et al. Microalgae as bioreactors for bioplastic production. Microbial Cell Factories. 2011 Dec;**10**(1):1-6

[73] Kamravamanesh D, Lackner M, Herwig C. Bioprocess engineering aspects of sustainable polyhydroxyalkanoate production in cyanobacteria. Bioengineering. 2018 Dec;**5**(4):111

[74] Abd El-malek F, Khairy H, Farag A, Omar S. The sustainability of microbial bioplastics, production and applications. International journal of biological macromolecules. 2020 Aug 15;**157**:319-328

[75] Narancic T, Cerrone F, Beagan N, O'Connor KE. Recent advances in bioplastics: application and biodegradation. Polymers. 2020 Apr;**12**(4):920

[76] Ummalyma SB, Sahoo D, Pandey A. Microalgalbiorefineries for industrial

products. InMicroalgae Cultivation for Biofuels Production 2020 Jan 1 (pp. 187-195). Academic Press.

[77] Alam MA, Xu JL, Wang Z, editors. Microalgae biotechnology for food, health and high value products. Singapore: Springer; 2020 Jan 22

[78] Rocha CJ, Álvarez-Castillo E, Yáñez MR, Bengoechea C, Guerrero A, Ledesma MT. Development of bioplastics from a microalgae consortium from wastewater. Journal of Environmental Management. 2020 Jun 1;**263**:110353

[79] Pathak S, Sneha CL, Mathew BB. Bioplastics: its timeline based scenario & challenges. Journal of Polymer and Biopolymer Physics Chemistry. 2014;**2**(4):84-90

[80] Bajhaiya AK, Dean AP, Zeef LA, Webster RE, Pittman JK. PSR1 is a global transcriptional regulator of phosphorus deficiency responses and carbon storage metabolism in Chlamydomonasreinhardtii. Plant Physiology. 2016 Mar;**170**(3):1216-1234

[81] Li DW, Balamurugan S, Yang YF, Zheng JW, Huang D, Zou LG, Yang WD, Liu JS, Guan Y, Li HY. Transcriptional regulation of microalgae for concurrent lipid overproduction and secretion. Science advances. 2019 Jan 1;5(1): eaau3795.

Chapter 6

Technological Advances in Synthetic Biology for Cellulosic Ethanol Production

Antonio Luiz Fantinel, Rogério Margis, Edson Talamini and Homero Dewes

Abstract

The resurgence of biofuels in the recent past has brought new perspectives for renewable energy sources. Gradually the optimistic scenarios were being challenged by the competition for raw materials dedicated to direct or indirect human food. Second-generation biorefineries have emerged as technological alternatives to produce biofuels from lignocellulosic biomass. The third generation of biorefineries uses alternative raw materials like algae and microalgae. Despite the technical feasibility, these biorefineries were indebted for their economic performance. Synthetic biology has provided new microbial platforms that are increasingly better adapted to industrial characteristics to produce biofuels and fine chemicals. Synthetic biology bioengineers microorganisms to take advantage of the low-cost and less-noble raw materials like lignocellulosic biomass, carbon dioxide, and waste as a sustainable alternative for bioenergy generation using bio-substrates. In this chapter, we analyze the innovations in synthetic biology as applied to cellulosic ethanol production based on registered patents issued over the last twenty years (1999–2019). Using Questel-Orbit Intelligence, we recovered a total of 298 patent families, from which we extracted the key concepts and technology clusters, the primary technological domains and applications, the geographical distribution of patents, and the leading patents assignees. Besides, we discuss the perspectives for future research and innovations and the market and policy opportunities for innovation in this technological field. We conclude that the patented technologies serve as a proxy for the development of synthetic biotechnology applied in cellulosic ethanol production by the fourth generation of biorefineries.

Keywords: Metabolic engineering, microorganisms, CRISPR, advanced biofuels, genetic engineering

1. Introduction

The transition from a fossil resource-based economy to a bio-based economy necessarily goes using synthetic biotechnologies [1–3]. Synthetic biology has been evolving and positively affecting human life by providing the opportunity to design and build new biological parts, devices, and systems that do not exist or redesign existing biological systems [4, 5] to produce biofuels and other chemicals [6, 7].

IntechOpen

The overlap between synthetic biology and bioeconomy occurs when we consider the latter part of the economy that uses new biological knowledge for commercial and industrial purposes, improving human well-being [8]. This perception intensifies when we ponder the sustainable use of biomass for non-food-biofuel production [9, 10].

Currently, ethanol produced from sugarcane in Brazil [11] and corn in the US [12] is the main alternative in the global supply chain of renewable fuels as a substitute for gasoline. However, this phenomenon fosters scientific debates about land use and food security, given that these are raw materials based on starch and sugar and that can be intended, directly or indirectly, for human food [13, 14].

In microbial ethanol production from lignocellulosic biomass, the dependence on food-related feedstocks is overcoming, being a sustainable alternative for bioenergy generation using substrates from the bioeconomy world [15, 16]. Lignocellulosic biomass is one of the most abundant feedstocks on the globe [17], with a production of approximately 181.5 billion tons/year [18], and can bring about significant changes in socioeconomic, agricultural, and energy systems when efficiently employed [19].

To overcome critical steps of microbial fermentation processes and to increase yields, technological advances are necessary on synthetic biology tools like metagenomics [20], genetic engineering [21], orthogonal communication systems [22], metaproteomics [23], metabolomics [24], and metabolic engineering [25–27].

From an industrial point of view, investment in these technological solutions depends upon the technical feasibility of using a particular organism to produce a specific compound and on the economic feasibility that results in profitable activity in the long run. Low yields in industrial processes are still recurrent due to low cell density, slowing down the efficient industrial expansion of this field [28–32] and are critical for biofuel production. Competitive synthetic biology technologies for ethanol production are estimated to be available in the coming years [33].

In the present chapter, we analyze the applications of synthetic biology tools related to cellulosic ethanol production from registered patents, visualizing the technological trends and their regional, institutional, and R&D markets distribution in the years 1999–2019. Patent analysis is one of the approaches to access innovative technologies and commercial aspects of a specific field [34]. The interest in searching for patents on cellulosic ethanol [35–38] or synthetic biology [39–41] is expanded here towards the technological development to future energy needs guided by the sustainable bioeconomy agenda. A total of 298 patent families were retrieved using Questel-Orbit Intelligence software. From them, we provide a high-quality dataset from the Questel-Orbit database that can contribute to formulating strategies and policies geared towards the development of these technologies and their applications in emerging markets, ensuring bioeconomic development for the next generations [33].

2. The evolution of innovations in synthetic biology and cellulosic ethanol

Figure 1 shows the distribution of the 298 synthetic biology patent families related to cellulosic ethanol throughout the twenty-year interval 1999–2019. The values correspond to the total frequencies following the International Patent Classification (IPC). In the eight-year interval 1999–2006, the annual number of published patents was negligible, with the first patent application occurring only in 2001. From 2007 to 2011, the number of applications on this topic showed

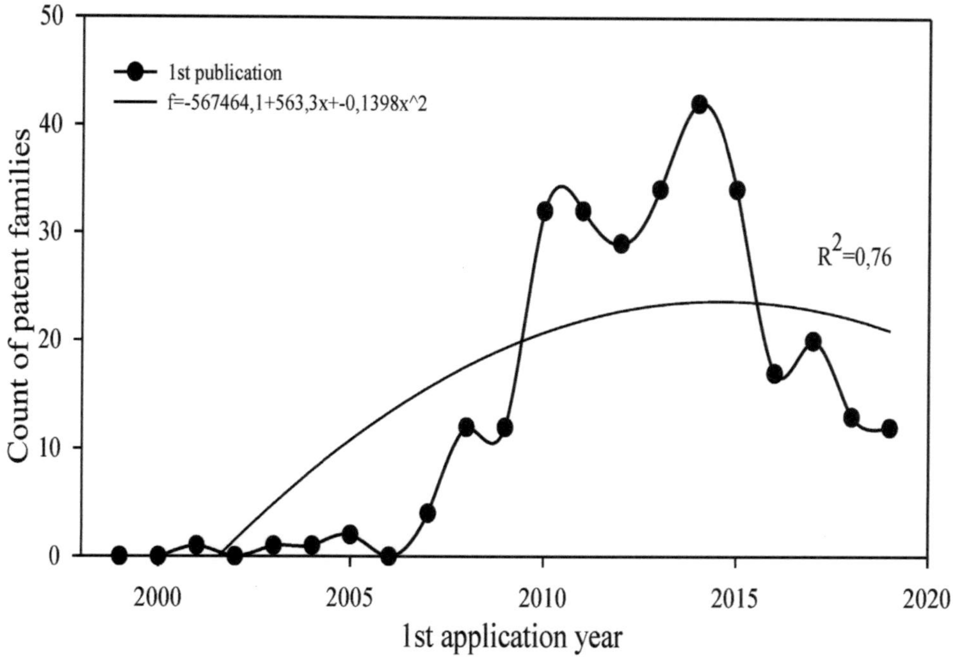

Figure 1.
Frequency of synthetic biology patents related to cellulosic ethanol. Source: Research data from Questel-orbit platform.

a considerable increase, peaking in 2010 and 2011. Subsequently, applications decreased in 2012, before temporarily recovering in 2013 and 2014. After that year, the applications decreased considerably, as observed through the trend line of synthetic biology patent applications for cellulosic ethanol. Thus, from 2015 there is a decline and subsequent stabilization, which we can understand as a transition from growth to maturity [42] or a consequence of time lag between patent application and patent grant.

2.1 Key concepts and technology clusters

The distribution of the main concepts among the retrieved patent families is presented in **Figure 2**. Nine semantic clusters regularly used by patent applicants were identified (**Figure 2A**). Most of these patents are related to the use of microorganisms (yeast and gram-negative bacteria), enzymatic activity, biomass, fermentation product, and biofuel production. As for the application of the technologies, we found the predominance of raw materials from biomass, such as agricultural residues (corn straw, wheat, rice, sugarcane bagasse, and switchgrass) and their main fermentable sugars (xylose, hemicellulose, and arabinose) for ethanol production. In the first years, the term "xylose" appeared more prominently than the others (**Figure 2B**), followed by "corn stover" and "lignocellulosic biomass". In the sequence, these terms were accompanied by the words "ethyl alcohol" given that the field includes this focus.

Therefore, by identifying the concepts commonly employed in the field of synthetic biology concerning cellulosic ethanol, we can propose insights for the development or identification of protected technologies in an emerging technological field with a view to its industrial application.

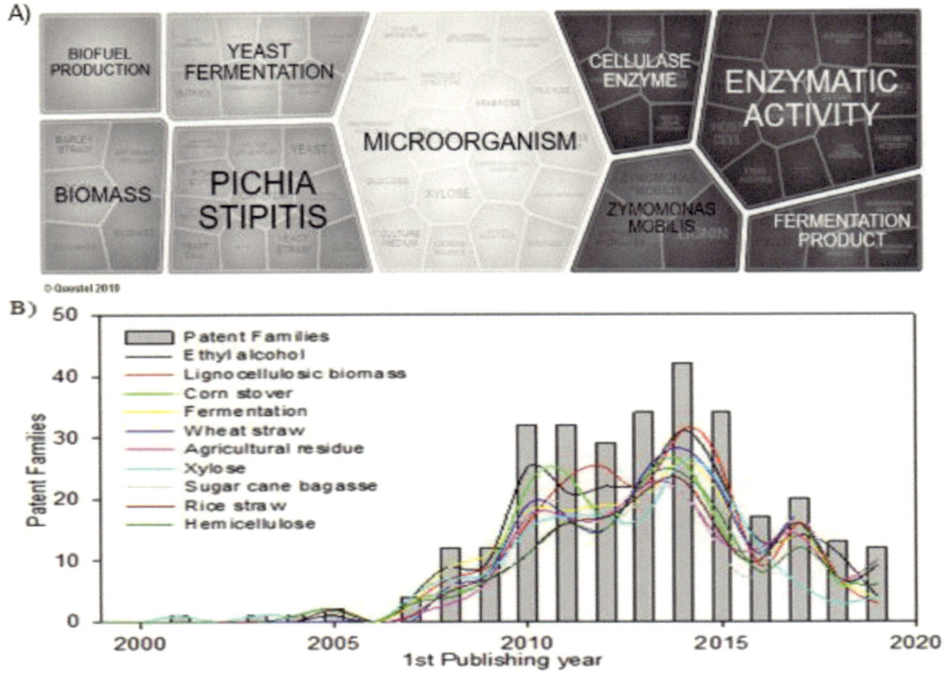

Figure 2.
Prevailing concepts in the retrieved patent families. 2A provides an overview of the content of this portfolio formed from the 298 synthetic biology patent families for cellulosic ethanol. 2B shows the distribution of these concepts concerning the complete portfolio over the twenty years surveyed. Source: Research data from Questel-orbit platform.

2.2 Major technological domains and applications

To elucidate the main technological domains and applications of the patent families, we analyzed the codes predominantly used to classify them (**Table 1**) following the structure proposed by IPC [43]. Approximately 40% of synthetic biology patent families related to cellulosic ethanol belong to the C12P classes. This class contemplates inventions concerning fermentation processes or using enzymes to synthesize a desired chemical composition or compound or to separate optical isomers of a racemic mixture. Within this class, group C12P-007, related to the preparation of organic compounds, represents 37% of the patents. Next in importance is class C12N (30% of the patent families), which deals with microorganisms or enzymes, or compositions thereof; propagation, preservation, or maintenance of microorganisms; genetic or mutation engineering; and culture media. Group C12N-001 accounts for 20% of the patents in this class, concerning processes for propagation, maintenance, or preservation of microorganisms or their compositions, or preparation, isolation of compositions containing a microorganism, and culture media for such.

When analyzed separately for each of the predominant codes (**Table 1**), we see the prominence of code C12P-007/06 (29 patent families). This code is related to the preparation of organic compounds containing oxygen, such as fuel ethanol, whose preponderant claims include yeast capable of fermenting xylose in the presence of glucose [44, 45], development of pretreated biomass [46, 47], use of yeast and bacteria in the presence of glycerol [48]. Among other technologies, we identified inventions related to methods for engineering *Thermoanaerobacterium saccharolyticum* [48], bioprocessing using recombinant *Clostridium* [48], methods for ethanol

IPC Codes	Code description	Family frequency
C12P-007/06	Preparation of oxygen-containing organic compounds [2006.01] • containing a hydroxy group [2006.01] •• acyclic [2006.01] ••• Ethanol, i.e. non-beverage [2006.01]	29
C12P-007/10	•••• produced as by-product or from waste or cellulosic material substrate [2006.01] ••••• substrate containing cellulosic material [2006.01]	20
C12N-001/21	• Bacteria; Culture media therefor [2006.01] •• modified by introduction of foreign genetic material [2006.01]	13
C12P-007/16	••• Butanols [2006.01]	13
C12N-001/20	• Bacteria; Culture media therefor [2006.01]	10
C12N-015/81	• Recombinant DNA-technology [2006.01] •• Introduction of foreign genetic material using vectors; Vectors; Use of hosts therefor; Regulation of expression [2006.01] ••• Vectors or expression systems specially adapted for eukaryotic hosts [2006.01] •••• for fungi [2006.01] ••••• for yeasts [2006.01]	10
C12P-007/18	••• polyhydric [2006.01]	9
C12P-007/14	•••• Multiple stages of fermentation; Multiple types of microorganisms or reuse for microorganisms [2006.01]	9
C12N-001/19	• Fungi (culture of mushrooms A01G 18/00; as new plants A01H 15/00); Culture media therefor [2006.01] •• Yeasts; Culture media therefor [2006.01] ••• modified by introduction of foreign genetic material [2006.01]	8
C12N-001/22	• Processes using, or culture media containing, cellulose or hydrolysates thereof [2006.01]	7

Source: research data from Questel-Orbit platform.

Table 1.
Leading IPC codes for synthetic biology patent families for cellulosic ethanol.

and hydrogen production using microorganisms [49], methods for propagation of microorganisms for hydrolysate fermentation [50], development of fermentation processes using transketolase/thiaminapyrophosphate enzymes [51], and hydrolysis of cellulosic material augmented with an enzyme composition [52].

Accordingly, code C12P-007/10 specifically addresses waste or cellulosic material or substrate containing cellulosic material for ethanol production and accounts for twenty (20) major patent families. These inventions relate to the development of microorganisms fermenting xylose and arabinose into ethanol [53], co-fermentation of pretreated lignocellulosic biomass [54, 55], wet oxidation methods of biomass [56], use of genetic engineering in microorganisms and enzymes [57].

The development of bacteria and culture mediums modified by the introduction of exogenous genetic material is classified by codes C12N-001/20 and C12N-001/21, which together represent twenty-three (23) major patent families. The inventions relate to the development and adaptation of *Zymomonas mobilis* strains [58], *C. thermocellum* [59, 60], *E. coli* [60], and anaerobic thermophilic bacteria [61, 62] for ethanol production. In addition, we verified the existence of technologies for the removal or inactivation of microbial inhibitors in biomass hydrolysates [63] and the conversion of xylose [64, 65] and arabinose [65] into ethanol.

Preparation of oxygen-containing organic compounds to produce butanol (C12P-007/16) features ten (10) patent families. The technologies pertinent to this code are related to recombinant microbial host cells of *S. cerevisiae* capable of converting hemicellulosic material into butanol-like alcohols [66], separation of undissolved solids after liquefaction [66] co-production of biofuels [66], microorganisms using protein and carbohydrate hydrolysates from biomass [67], and genetic engineering in bacteria [67] and yeast [68].

The use of recombinant DNA technologies via vectors and expression regulation in yeast and fungi categorized by code C12N-015/81 presents ten (10) main patent families. These patents target the development of yeast cells with xylose isomerase activity [69], culture medium and bioreactors [70], L-arabinose transporter polypeptide (I) from *Pichia stipitis* [70], gene transcription control [70], glycerol-free ethanol production using recombinant yeast [71, 72], microbial cells capable of transporting xylo-oligosaccharides [72], and yeast cells with a reduced enzyme activity for NADH-dependent glycerol synthesis [72].

The preparation of organic compounds containing at least two hydroxyl groups (C12P-007/18) features nine (9) main patent families. In this code, inventions are directed to the development of non-native pentose metabolic pathways in yeast cells [73], yeast genes encoding enzymes in the pentose pathway [74], genetically modified thermophilic or mesophilic microorganism [74], *S. cerevisiae* strains with reduced glycerol productivity [74] and fermentation microorganism propagation [75].

New forms of fermentation through multiple stages, different types of microorganisms, or reuse of microorganisms represented by code C12P-007/14 present nine (9) main patent families. The technologies are related to the production of syrups enriched with C5 and C6 sugars [75], ethanol production from lignocellulosic biomass [76] and xylitol production from biomass with enriched pentose component [77]. Methods for pectin degradation [78], pretreated cellulosic material [78], biocatalyst development [79] and microorganism propagation [80] for ethanol production are also checked in this code.

Yeast modification by introducing exogenous genetic material represented by C12N-001/19 features eight (8) main patent families. The inventions relate to the use of metabolic engineering for the elimination of the glycerol pathway [78], joint utilization of xylose and glucose [78], and rapid fermentation of xylose [78] in yeast. Methods for enhanced expression of a glycolytic system enzyme [78], glycerol transport [81] and alpha-ketoisovalerate conversion to isobutyraldehyde [81] also integrate this code.

The tenth code with seven (7) patent families relates to processes using culture medium containing cellulose or hydrolysates (C12N-001/22). The inventions concern continuous xylose growth using *Zymomonas* [81], oligosaccharide degradation by recombinant host cells [82] and lignocellulose bioprocessing employing recombinant *Clostridium* [83]. Methods for glycerol reduction in biomass fermentative processes [83], increasing tolerance to acetate toxicity in recombinant microbial host cells [84] and controlling contamination during fermentation [84] also integrate this code.

The knowledge present in these technological domains and their applications allows researchers to identify potential fields of development of new cellulosic ethanol production routes using synthetic biology as a technical platform.

2.3 The geographical distribution of innovations

Next, we analyzed the geographical distribution of synthetic biology patent families related to cellulosic ethanol, according to priority country (**Figure 3**).

We found only 14 priority countries holding this technology. The lead-in technological innovation in this field is the USA (**Figure 3**) since approximately 67% of the patents recovered are in the name of American applicants. The main technological applications patented by American inventors are focused on the production of ethanol from biomass by-products or wastes (C12P-007/10; C12P-007/06), as well as modification of bacteria (C12N-001/21) and fungi (C12N-001/19) by introducing endogenous genetic material as applications to overcome the current barriers to the conversion of biomass to ethanol [85, 86]. Following at a distance is the European Patent Organization, followed by Japan and China, which account for 13%, 6%, and 3% of patent families, respectively. The remaining patent families, which total 11%, are distributed among ten other countries.

The global distribution of patent families protected in the various offices can be seen in **Figure 4**. The data corroborates the identification of target markets and demonstrates the patenting strategies of the applicant countries. The illustration confirms that demand is concentrated in the United States, with 49% of patent families, followed by the European Patent Organization (37% of families), India (33% of families), Brazil (30% of families), and China (29% of patent families).

Through this data, the strategies for patent protection used by applicants in the sector studied are identified. The preference for registration in patent offices in certain countries indicates the potential of the markets from the viewpoint of the need for commercial protection of new industrial technologies.

2.4 Leading patent assignees

The main assignees of patents in synthetic biology associated with cellulosic ethanol production encompass both private companies and educational and research institutions. The applicants were analyzed by the number of active patents, the average size of these families, generality index, and originality (**Table 2**). The number of patents and the average size of patent families refer to active patents and their breadth, respectively. In turn, the generality index is defined by Hall et al. [87] as the range of fields of future citations of a given patent. Future citations can be used to assess the subsequent generations of an invention that have benefited from an issued patent by measuring the range of technology fields and, consequently, of industries that cite that patent [88, 89]. On the other hand, the originality index measures the range of technological fields in which a patent is based [89, 90].

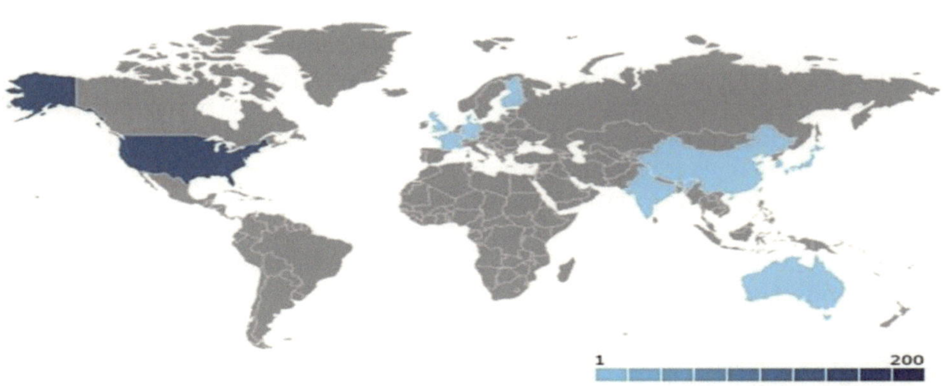

Figure 3.
Distribution of priority patent applications in the various offices over the last 20 years (1999–2019). Source: Research data from Questel-orbit platform.

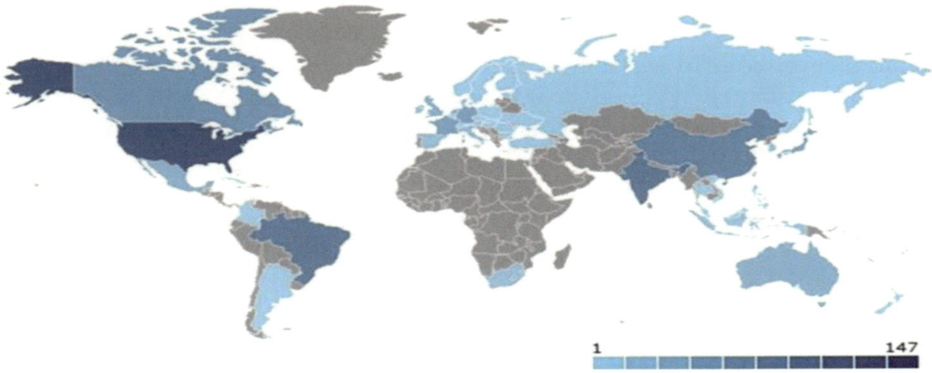

Figure 4.
Worldwide distribution of patents under protection by national patent offices over the last 20 years (1999–2019). Source: Research data from Questel-orbit platform.

Assignees	Active patent families	Average family size	Indicators	
			Originality	Generality
Novozymes	27	5,8	0,91	0,87
Du Pont De Nemours	18	8,7	0,87	0,87
Butamax Advanced Biofuels	17	9,3	0,90	0,88
Lallemand	14	5,7	0,85	0,84
DSM	11	6,5	0,84	0,81
Danisco	10	8,6	0,90	0,88
University of Florida	6	1,7	0,84	0,87
Toray Industries	9	7,1	0,92	0,84
DSM Ip Assets	7	7,4	0,83	0,78
University of California	3	2,7	0,86	0,88

Source: research data from Questel-Orbit platform.

Table 2.
Patent families by assignees and value indicators.

Novozymes, the largest patent holder (27 patent families), is a Danish company that develops and markets enzymes for industrial use. We also highlight the American companies DuPont De Nemours (18 families of patents) and Butamax Advanced Biofuels (17 families). Butamax emerged from the partnership of DuPont and BP, so that in 2017, it acquired the company Nesika Energy LLC, installing an ethanol production plant in Scandia County in the state of Kansas-US, to add to this unit the production of bio-isobutanol. The top five global companies holding patents on the analyzed technology include Canadian Lallemand with 14 patent families and the Dutch company DSM, which has 12 patent families. Regarding the average family size of patents, Butamax Advanced Biofuels is configured with the largest average family size, about 9.3, followed by DuPont De Nemours (8.7) and Danisco (8.6).

Butamax Advanced Biofuels (0.88), Danisco (0.88), the University of California (0.88), Novozymes (0.87), and Du Pont De Nemours (0.87) have the highest patent generality indices and, consequently, tend to account for the most relevant applications. Toray Industries (0.92) and Novozymes (0.91) show the highest scores for the originality index. The importance of the companies cited for inventions and subsequent innovations in the technological field analyzed is undeniable.

We emphasize that, except for Butamax Advanced Biofuels that aims at the production and commercialization of bio-isobutanol, the other companies aim at developing and commercializing enzymes, yeasts, and catalysts for the production of advanced biofuels.

3. Perspectives for future research and innovations

The present study gathers evidence of technological opportunities for ethanol production from raw materials derived from the bioeconomy. As evidenced in our findings, the development of innovations in this field requires multidisciplinary knowledge, providing solutions for industrial applications, which employ *S. cerevisiae*, *E. coli*, and *Z. mobilis* [91]. However, these potentially usable microorganisms in these fermentative processes are not naturally adaptable to extreme industrial conditions [92] or do not tolerate high concentrations of inhibitory compounds released during biomass fermentation [93]. Thus, to overcome these barriers, different synthetic biology and metabolic engineering approaches are employed to microorganisms to make them robust living factories adapted to the industrial activities required for biomass fermentation into ethanol [5, 94, 95]. These insights about synthetic biology may allow folding and probing the genome at different length and time scales, making it possible to understand gene positioning and functions [96]. Nevertheless, we check the prospect of new unconventional yeasts and bacteria such as *P. stipitis* for fermentation of lignocellulosic biomass.

Because the yeast *S. cerevisiae*, commonly used in ethanol fermentation of sugar-based feedstocks, is not a natural degrader of arabinose [97] and xylose [98, 99], making the fermentation processes accessible to these sugars requires pathway engineering [100, 101]. Ye et al. [102] integrated a heterologous fungal arabinose pathway into *S. cerevisiae*, with the deletion of the PHO13 phosphatase gene, increasing the rate of arabinose consumption and ethanol production under aerobic conditions. In Cunha et al. [103], two pathways (XR/XDH or XI) of xylose assimilation by *S. cerevisiae* were compared in ethanol production under different fermentation conditions, demonstrating satisfactory results for the feasibility of this fuel from non-detoxified hemicellulosic hydrolysates. Meanwhile, Mitsui et al. [104] developed a novel genome shuffling method using CRISPR-Cas to improve stress tolerance in S. cerevisiae yeast. Regarding *E. coli*, its main disadvantages refer to the narrow growth range of neutral pH (6.0–8.0), in addition to ethanol not being a core product for this bacterium. However, Sun et al. [105] successfully developed an efficient bioprocess using an *E. coli* strain for ethanol production and xylose recovery from corn cob hydrolysate. Strains of this bacterium with regulated glucose utilization showed efficient metabolism of mixed sugars in lignocellulosic hydrolysates, and higher ethanol production yields [106]. In the same perspective, metabolic engineering has been studied to provide simultaneous utilization of glucose and xylose in this bacterial culture [107].

High cellulosic ethanol yields are achieved using *Z. mobilis* strains due to their unique physiology [108, 109]. It is possible to employ other substrates, mitigating the socio-environmental challenges for expanding ethanol production [30, 110]. Different approaches have been tested in *Z. mobilis* to improve the fermentation of lignocellulosic biomass substrates into ethanol [111, 112].

One critical step in developing methods of the microbial fermentation process of lignocellulosic biomass is its pre-treatment to increase the digestibility of the available sugars. Lignocellulosic biomass consists of highly crystalline cellulose and a hemicellulose sheath wrapped in a lignin network. This structure causes recalcitrance in fermentation processes [113, 114]. Recalcitrance is the main obstacle to using lignocellulosic biomass for ethanol production. It determines the rest of the fermentation and the overall efficiency of the process [115]. Biological pre-treatment has been employed for the deconstruction of this biomass because of its wide application, lower energy consumption, no generation of toxic substances, and higher yield [116].

Co-fermentation of different sugars from lignocellulosic biomass and its residues enables ethanol production processes to become economically viable [117]. The potential of using a blend of *E. coli* strains and yeasts to rapidly ferment all sugars in pretreated biomass at high ethanol rates is presented by Wang et al. [118]. In the same perspective, Amoah et al. [119] developed a yeast with xylose assimilation capable of co-fermenting xylose and glucose in ionic liquid for ethanol production from lignocellulosic biomass. Advances are also in attempting to overcome obstacles and perturbations present in the degradation of lignin [120] and cellulose [121] using microbial consortium and genetic engineering via RNA-guided Cas9 in *S. cerevisiae* [21], *Candida glycerinogenes* [122], *Rhodosporidium toruloides* [123]. The results denote significant increases in stress tolerance of microorganisms in severe fermentative processes.

4. Market and policy opportunities for innovation

The continued development of synthetic biology R&D for cellulosic ethanol production depends on both the technical and economic feasibility of the solutions presented. In the analyzed period, approximately US$ 820 million was invested in synthetic biology research aimed at the development of advanced biofuels and bioproducts from microbial systems [124].

Despite being the second-largest ethanol producer in the world, Brazil does not own priority patents registered in synthetic biology for cellulosic ethanol production, becoming only a target market for other countries holding these technologies, like the USA. In Brazil, the unit cost per protected patent is very high, corresponding to approximately US$ 13,000 per patent. Besides the poor institutional environment for innovation, the unit cost of protection may be one of the reasons for the lack of patent applications by Brazilian assignees. According to Cicogna et al. [125], Brazil is an example of an infant industry that is slowly reaching maturity. In the same perspective, Kang et al. [126] point to the need for government policies that facilitate the development of promising renewable technologies, in addition to offering incentives for their commercialization.

In an attempt to change this situation, in 2017 a new national policy for biofuels was enacted by the Brazilian government, the RenovaBio, aiming to promote ethanol and biodiesel production from various sources available in the country [127]. Brazil, in the future, could become the largest producer of bio-based products when economic, logistical, regulatory, and political barriers are overcome [128]. Its territorial extension and diverse regional edaphoclimatic characteristics enable the country an intensive production of biomass for industrial biotechnology at a relatively lower cost compared to other locations that prospect synthetic microbial cells. Moreover, it is one of the world's leading food producers with agroindustry generating a significant amount of waste with potential for transformation into bioenergy, providing a new pathway for biofuel production not competing with food but biomass and agricultural waste [129].

The knowledge applied to the creation of new technologies in synthetic biology related to cellulosic ethanol comes mainly from companies that work in the development of enzymes and microorganisms for the transformation of biomass into ethanol and also in the production and commercialization of this biofuel. Companies seek, through patents, the commercial exploitation of these new technologies as they maximize their competitive advantages [130]. Industries operate in complex technological environments. Their technical knowledge is highly relevant to gain a competitive advantage. Therefore, companies cannot rely solely on their internal R&D units but also need to seek support from external sources of technology. To protect their inventions from third-party misuse, innovative companies seek patent protection [131, 132].

5. Concluding remarks

In this chapter, we examined the developments and applications of synthetic biology tools related to cellulosic ethanol by analyzing patents to investigate the current stage and dynamics of this technological field and its role as a proxy for a sustainable bioeconomy using non-food feedstocks. The findings are not necessarily only involved in the field of synthetic biology, but also in its numerous approaches that could circumscribe the development of cellulosic ethanol production worldwide. Our analyses provide a compilation of relevant patents, allowing us to understand, track, and project the role of synthetic biology in fostering solutions for the emerging sustainable bioeconomy, and enabling socio-market scenarios with this orientation.

Thinking about sustainable bioeconomy for energy generation, the use of synthetic biology tools may provide new living factories increasingly adapted to industrial processing technology, despite the decrease in the search for patent applications. Using the results from this study, synthetic or bioenergy engineers will be able to choose robust microorganisms capable of performing optimized fermentation processes or biomass processing methods, alleviating a bottleneck that limits the yields of bioenergy research. As these efforts mature, they can be expanded into biofuel production based on bioeconomic-nonfood substrates.

Overall, the research has provided approach for evaluating synthetic biology R&D performance related to cellulosic ethanol and bioeconomy. The results can help researchers quickly integrate into the field as they will easily understand the technological frontiers. The study also provides references for future energy research and policies that could proxy for a world focused on a more sustainable bioeconomy using non-food feedstocks. In addition, the text illustrated the importance of knowledge spillovers in R&D and signaled possibilities for future work. Deepening the understanding of cellular systems can raise the yield of low-cost carbon sources for cellulosic ethanol production. Integration of different generations of technologies may be an alternative to improve the total yields and make cellulosic ethanol economically viable.

Acknowledgements

A.L.F., R.M., and E.T. are grateful for supporting received from the Conselho Nacional de Desenvolvimento Científico e Tecnológico – CNPq, and Coordenação de Aperfeiçoamento de Pessoal de Nível Superior (CAPES). This study was carried out with financial fellowship supports 309586/2019-4, 303956/2019-4, and 141259/2016-7 from the Conselho Nacional de Desenvolvimento Científico e Tecnológico – CNPq.

Conflict of interest

The authors inform they have no conflict of interest to declare.

Author details

Antonio Luiz Fantinel[1*], Rogério Margis[2], Edson Talamini[1,3] and Homero Dewes[1,2]

1 Interdisciplinary Center for Studies and Research in Agribusiness - CEPAN, Universidade Federal do Rio grande do Sul - UFRGS, Porto Alegre, RS, Brazil

2 Department of Biophysics, Universidade Federal do Rio grande do Sul - UFRGS, Porto Alegre, RS, Brazil

3 Department of Economics and International Relations - DERI, Faculty of Economics - FCE, Porto Alegre, RS, Brazil

*Address all correspondence to: tonifantinel@gmail.com

References

[1] de Lorenzo, V, Schmidt, M. Biological standards for the Knowledge-Based BioEconomy: What is at stake. New Biotechnology. 2018; 40:170-180. DOI: 10.1016/j.nbt.2017.05.001

[2] French, KE. Harnessing synthetic biology for sustainable development. Nature Sustainability. 2019;2(4):250-252. DOI: 10.1038/s41893-019-0270-x

[3] Srivastava, A, Villalobos, MB, Singh, RK. Engineering Photosynthetic Microbes for Sustainable Bioenergy Production, in: Contemporary Environmental Issues and Challenges in Era of Climate Change; 2020. pp. 183-198. DOI: 10.1007/978-981-32-9595-7_10

[4] Benner, SA, Sismour, AM. Synthetic biology. Nature Reviews Genetics. 2005;6(7):533-543. DOI: 10.1038/nrg1637

[5] Ren, J, Lee, J, Na, D. Recent advances in genetic engineering tools based on synthetic biology. Journal of Microbiology. 2020;58(1):1-10. DOI: 10.1007/s12275-020-9334-x

[6] Nielsen, J, Keasling, JD. Synergies between synthetic biology and metabolic engineering. Nature Biotechnology. 2011;29(8):693-695. DOI: 10.1038/nbt.1937

[7] Chownk, M, Thakur, K, Purohit, A, Vashisht, A, Kumar, S. Applications and Future Perspectives of Synthetic Biology Systems, in: Current Developments in Biotechnology and Bioengineering; 2019. P. 393-412. DOI: 10.1016/b978-0-444-64085-7.00016-2

[8] Zilberman, D, Kim, E, Kirschner, S, Kaplan, S, Reeves, J. Technology and the future bioeconomy. Agricultural Economics (United Kingdom). 2013;44(s1):95-102. DOI: 10.1111/agec.12054

[9] Aro, EM. From first generation biofuels to advanced solar biofuels. Ambio. 2016;45(1):24-31. DOI: 10.1007/s13280-015-0730-0

[10] Dupont-Inglis, J, Borg, A. Destination bioeconomy – The path towards a smarter, more sustainable future. New Biotechnology. 2018;40:140-143. DOI: 10.1016/j.nbt.2017.05.010

[11] Rudorff, BFT, de Aguiar, DA, da Silva, WF, Sugawara, LM, Adami, M, Moreira, MA. Studies on the rapid expansion of sugarcane for ethanol production in São Paulo state (Brazil) using Landsat data. Remote Sensing. 2010;2(4): 1057-1076. DOI: 10.3390/rs2041057

[12] Pimentel, D, Patzek, TW. Ethanol production using corn, switchgrass, and wood; biodiesel production using soybean and sunflower, in: Food, Energy, and Society; 2017. DOI: 10.1201/9781420046687

[13] Tomei, J, Helliwell, R. Food versus fuel? Going beyond biofuels. Land Use Policy. 2016;56:320-326. DOI: 10.1016/j.landusepol.2015.11.015

[14] Rulli, MC, Bellomi, D, Cazzoli, A, De Carolis, G, D'Odorico, P. The water-land-food nexus of first-generation biofuels. Scientific Reports. 2016;6:22521. DOI: 10.1038/srep22521

[15] Guerriero, G, Hausman, JF, Strauss, J, Ertan, H, Siddiqui, KS. Lignocellulosic biomass: Biosynthesis, degradation, and industrial utilization. Engineering in Life Sciences. 2016;16(1);1-16. DOI: 10.1002/elsc.201400196

[16] Lara-Flores, AA, Araújo, RG, Rodríguez-Jasso, RM, Aguedo, M, Aguilar, CN, Trajano, HL, Ruiz, HA. Bioeconomy and Biorefinery: Valorization of Hemicellulose from

Lignocellulosic Biomass and Potential Use of Avocado Residues as a Promising Resource of Bioproducts. In Waste to Wealth; 2016. p. 141-170. Springer, Singapore. DOI: 10.1007/978-981-10-7431-8_8

[17] Narnoliya, LK, Jadaun, JS, Singh, SP. Management of Agro-industrial Wastes with the Aid of Synthetic Biology. In Biosynthetic Technology and Environmental Challenges; 2018. p. 11-28. DOI: 10.1007/978-981-10-7434-9_2

[18] Paul, S, Dutta, A. Challenges and opportunities of lignocellulosic biomass for anaerobic digestion. Resources, Conservation and Recycling. 2018;130:164-174. DOI: 10.1016/j.resconrec.2017.12.005

[19] Ingle, AP, Ingle, P, Gupta, I, Rai, M. Socioeconomic impacts of biofuel production from lignocellulosic biomass, in: Sustainable Bioenergy, Elsevier; 2019. p. 347-366. DOI: 10.1016/b978-0-12-817654-2.00013-7

[20] Chen, Y, Wang, Y, Chen, TH, Yao, MD, Xiao, WH, Li, BZ, Yuan, YJ. Identification and manipulation of a novel locus to improve cell tolerance to short-chain alcohols in *Escherichia coli*. Journal of Industrial Microbiology & Biotechnology. 2018;45:589-598. DOI: 10.1007/s10295-017-1996-y

[21] EauClaire, SF, Zhang, J, Rivera, CG, Huang, LL. Combinatorial metabolic pathway assembly in the yeast genome with RNA-guided Cas9. Journal of Industrial Microbiology and Biotechnology. 2016;43(7):1001-1015. DOI: 10.1007/s10295-016-1776-0

[22] Bervoets, I, Van Brempt, M, Van Nerom, K, Van Hove, B, Maertens, J, De Mey, M, Charlier, D. A sigma factor toolbox for orthogonal gene expression in *Escherichia coli*. Nucleic Acids Research. 2018;46:2133-2144. DOI: 10.1093/nar/gky010

[23] Speda, J, Jonsson, BH, Carlsson, U, Karlsson, M. Metaproteomics-guided selection of targeted enzymes for bioprospecting of mixed microbial communities. Biotechnology for Biofuels. 2017;10(1):128. DOI: 10.1186/s13068-017-0815-z

[24] Pandey, S. Prospects of metagenomic cellulases for converting lignocellulosic biomass into bio-ethanol. Journal of Pure and Applied Microbiology. 2017;11(2):1079-1091. DOI: 10.22207/JPAM.11.2.51

[25] Hollinshead, W, He, L, Tang, YJ. Biofuel production: An odyssey from metabolic engineering to fermentation scale-up. Frontiers in Microbiology. 2014;5: 344. DOI: 10.3389/fmicb.2014.00344

[26] Liao, JC, Mi, L, Pontrelli, S, Luo, S. Fuelling the future: Microbial engineering for the production of sustainable biofuels. Nature Reviews Microbiology. 2016;14(5):288. DOI: 10.1038/nrmicro.2016.32

[27] Duan, G, Wu, B, Qin, H, Wang, W, Tan, Q, Dai, Y, Qin, Y, Tan, F, Hu, G, He, M. Replacing water and nutrients for ethanol production by ARTP derived biogas slurry tolerant *Zymomonas mobilis* strain. Biotechnology for Biofuels. 2019;12(1): 124. DOI: 10.1186/s13068-019-1463-2

[28] Chisti, Y. Constraints to commercialization of algal fuels. Journal of Biotechnology. 2013;167:201-214. DOI: 10.1016/j.jbiotec.2013.07.020

[29] Rahman, Z, Rashid, N, Nawab, J, Ilyas, M, Sung, BH, Kim, SC. *Escherichia coli* as a fatty acid and biodiesel factory: current challenges and future directions. Environmental Science and Pollution Research. 2013;23:12007-12018. DOI: 10.1007/s11356-016-6367-0

[30] Khetkorn, W, Rastogi, RP, Incharoensakdi, A, Lindblad, P,

Madamwar, D, Pandey, A, Larroche, C. Microalgal hydrogen production – A review. Bioresource Technology. 2017;243:1194-1206. DOI: 10.1016/j.biortech.2017.07.085

[31] Xue, C, Zhao, JB, Chen, LJ, Yang, ST, Bai, FW. Recent advances and state-of-the-art strategies in strain and process engineering for biobutanol production by *Clostridium acetobutylicum*. Biotechnology Advances. 2017;35:310-322. DOI: 10.1016/j.biotechadv.2017.01.007

[32] Jullesson, D, David, F, Pfleger, B, Nielsen, J. Impact of synthetic biology and metabolic engineering on industrial production of fine chemicals. Biotechnology Advances. 2015;33(7): 1395-1402. DOI: 10.1016/j.biotechadv.2015.02.011

[33] Dehghani Madvar, M, Aslani, A, Ahmadi, MH, Karbalaie Ghomi, NS. Current status and future forecasting of biofuels technology development. International Journal of Energy Research. 2019;43(3):1142-116. DOI: 10.1002/er.4344

[34] Devarapalli, P, Deshpande, N, Hirwani, RR. Xylose utilization in ethanol production: a patent landscape. Biofuels, Bioproducts and Biorefining. 2016;10(5):534-541. DOI: 10.1002/bbb.1664

[35] Gustafsson, R, Kuusi, O; Meyer, M. Examining open-endedness of expectations in emerging technological fields: The case of cellulosic ethanol. Technological Forecasting and Social Change. 2015;91:179-193. DOI: 10.1016/j.techfore.2014.02.008

[36] Kessler, J, Sperling, D. Tracking U.S. biofuel innovation through patents. Energy Policy. 2016;98:97-107. DOI: 10.1016/j.enpol.2016.08.021

[37] Karvonen, M, Klemola, K. Identifying bioethanol technology generations from the patent data. World Patent Information. 2019;57:25-34. DOI: 10.1016/j.wpi.2019.03.004

[38] Toivanen, H, Novotny, M. The emergence of patent races in lignocellulosic biofuels, 2002-2015. Renewable and Sustainable Energy Reviews. 2017;77:318-326. DOI: 10.1016/j.rser.2017.03.089

[39] Van Doren, D, Koenigstein, S, Reiss, T. The development of synthetic biology: A patent analysis. Systems and Synthetic Biology. 2013;7(4):209-220. DOI: 10.1007/s11693-013-9121-7

[40] Carbonell, P, Gök, A, Shapira, P, Faulon, JL. Mapping the patent landscape of synthetic biology for fine chemical production pathways. Microbial Biotechnology. 2016;135:544-554. DOI: 10.1111/1751-7915.12401

[41] Ribeiro, B, Shapira, P. Private and public values of innovation: A patent analysis of synthetic biology. Research Policy. 2020;49(1):103875. DOI: 10.1016/j.respol.2019.103875

[42] Haupt, R, Kloyer, M, Lange, M. Patent indicators for the technology life cycle development. Research Policy. 2007;36(3):387-398. DOI: 10.1016/j.respol.2006.12.004

[43] WIPO. Guide to the International Patent Classification. World Intelectual Property Organization. 2019.

[44] WO2009093630A1 - Hexose-pentose cofermenting yeast having excellent xylose fermentability and method of highly efficiently producing ethanol using the same - Google Patents [WWW Document], n.d. URL https://patents.google.com/patent/WO2009093630A1/en?oq=WO2009093630 (accessed 1.17.20).

[45] US10301653B2 - Microorganisms that co-consume glucose with non-glucose carbohydrates and methods of

use - Google Patents [WWW Document], n.d. URL https://patents.google.com/patent/US10301653B2/en?oq=US10301653 (accessed 1.17.20).

[46] WO2009137804A1 - Yeast cells and mehtods for increasing ethanol production - Google Patents [WWW Document], n.d. URL https://patents.google.com/patent/WO2009137804A1/en?oq=WO2009%2F137804 (accessed 1.17.20).

[47] WO2015142399A1 - Preparation of biomass - Google Patents [WWW Document], n.d. URL https://patents.google.com/patent/WO2015142399A1/en?oq=WO2015%2F142399 (accessed 1.17.20).

[48] WO2012067510A1 - Yeast strains engineered to produce ethanol from glycerol - Google Patents [WWW Document], n.d. URL https://patents.google.com/patent/WO2012067510A1/en?oq=WO2012%2F067510 (accessed 1.17.20).

[49] WO2009124321A1 - Methods and compositions for improving the production of fuels in microorganisms - Google Patents [WWW Document], n.d. URL https://patents.google.com/patent/WO2009124321A1/en?oq=WO2009%2F124321 (accessed 1.17.20).

[50] WO2017112475A1 - Methods, systems, and compositions for propagation of a fermentation microorganism - Google Patents [WWW Document], n.d. URL https://patents.google.com/patent/WO2017112475A1/en?oq=WO2017%2F112475 (accessed 1.17.20).

[51] WO2010014817A2 - Producing fermentation products - Google Patents [WWW Document], n.d. URL https://patents.google.com/patent/WO2010014817A2/en?oq=WO2010%2F014817 (accessed 1.17.20).

[52] WO2010080408A2 - Methods for increasing enzymatic hydrolysis of cellulosic material in the presence of a peroxidase - Google Patents [WWW Document], n.d. URL https://patents.google.com/patent/WO2010080408A2/en?oq=WO2010%2F080408 (accessed 1.17.20).

[53] WO2013071112A1 - A genetically modified strain of s. cerevisiae engineered to ferment xylose and arabinose - Google Patents [WWW Document], n.d. URL https://patents.google.com/patent/WO2013071112A1/en?oq=WO2013%2F071112 (accessed 1.17.20).

[54] WO2014160402A1 - Co-conversion of carbohydrates to fermentation products in a single fermentation step - Google Patents [WWW Document], n.d. URL https://patents.google.com/patent/WO2014160402A1/en?oq=WO2014%2F160402 (accessed 1.17.20).

[55] WO2008124162A2 - Substrate-selective co-fermentation process - Google Patents [WWW Document], n.d. URL https://patents.google.com/patent/WO2008124162A2/en?oq=WO2008%2F124162 (accessed 1.17.20).

[56] WO2017049394A1 - Oxydation par voie humide de biomasse - Google Patents [WWW Document], n.d. URL https://patents.google.com/patent/WO2017049394A1/fr?oq=WO2017%2F049394 (accessed 1.17.20).

[57] WO2011150318A1 - Acetate-resistant yeast strain for the production of a fermentation product - Google Patents [WWW Document], n.d. URL https://patents.google.com/patent/WO2011150318A1/en?oq=WO2011%2F150318 (accessed 1.17.20).

[58] BR112013014912A2 - micro-organismo recombinante, zymomonas

recombinante, polipeptídeo, métodos para a produção de uma zymomonas recombinantes produtora de etanol e para a produção de etanol - Google Patents [WWW Document], n.d. URL https://patents.google.com/patent/BR112013014912A2/pt?oq=BR112013014912 (accessed 1.17.20).

[59] WO2014182054A1 - Recombinant microorganism metabolizing 3,6-anhydride-l-galactose and a use thereof - Google Patents [WWW Document], n.d. URL https://patents.google.com/patent/WO2014182054A1/en?oq=WO2014%2F182054 (accessed 1.17.20).

[60] WO2011137401A2 - Redirected bioenergetics in recombinant cellulolytic clostridium microorganisms - Google Patents [WWW Document], n.d. URL https://patents.google.com/patent/WO2011137401A2/en?oq=WO2011%2F137401 (accessed 1.17.20).

[61] BRPI0712490A2 - cepa bg1 de trehmoanaerobacter mathranii - Google Patents [WWW Document], n.d. URL https://patents.google.com/patent/BRPI0712490A2/pt?oq=BRPI0712490 (accessed 1.17.20).

[62] WO2010031793A2 - Thermophilic fermentative bacterium producing butanol and/or hydrogen from glycerol - Google Patents [WWW Document], n.d. URL https://patents.google.com/patent/WO2010031793A2/en?oq=WO2010%2F031793 (accessed 1.17.20).

[63] WO2013019822A1 - Removal of microbial fermentation inhibitors from cellulosic hydrolysates or other inhibitor-containing compositions - Google Patents [WWW Document], n.d. URL https://patents.google.com/patent/WO2013019822A1/en?oq=WO2013%2F019822 (accessed 1.17.20).

[64] BR112014015565A2 - método de produzir uma célula, célula e método para produzir etanol - Google Patents [WWW Document], n.d. URL https://patents.google.com/patent/BR112014015565A2/pt?oq=BR112014015565 (accessed 1.17.20).

[65] BR112014015566A2 - célula hospedeira bacteriana recombinante e processo para produção de etanol - Google Patents [WWW Document], n.d. URL https://patents.google.com/patent/BR112014015566A2/pt?oq=BR112014015566 (accessed 1.17.20).

[66] WO2010059616A2 - Biocatalysts and methods for conversion of hemicellulose hydrolsates to biobased products - Google Patents [WWW Document], n.d. URL https://patents.google.com/patent/WO2010059616A2/en?oq=WO2010%2F059616 (accessed 1.17.20).

[67] WO2014047421A1 - Production of renewable hydrocarbon compositions - Google Patents [WWW Document], n.d. URL https://patents.google.com/patent/WO2014047421A1/en?oq=WO2014%2F047421 (accessed 1.17.20).

[68] WO2015103001A1 - Expression of a hap transcriptional complex subunit - Google Patents [WWW Document], n.d. URL https://patents.google.com/patent/WO2015103001A1/en?oq=WO2015%2F103001 (accessed 1.17.20).

[69] WO2014098939A1 - Expression of xylose isomerase activity in yeast - Google Patents [WWW Document], n.d. URL https://patents.google.com/patent/WO2014098939A1/en?oq=WO2014%2F098939 (accessed 1.17.20).

[70] WO2011123715A1 - Metabolically engineered yeasts for the production of ethanol and other products from xylose and cellobiose - Google Patents [WWW Document], n.d. URL https://patents.

google.com/patent/WO2011123715A1/en?oq=WO2011%2F123715 (accessed 1.17.20).

[71] WO2019063542A1 - Improved glycerol free ethanol production - Google Patents [WWW Document], n.d. URL https://patents.google.com/patent/WO2019063542A1/en?oq=WO2019%2F063542 (accessed 1.17.20).

[72] WO2019063543A1 - Improved glycerol free ethanol production - Google Patents [WWW Document], n.d. URL https://patents.google.com/patent/WO2019063543A1/en?oq=WO2019%2F063543 (accessed 1.17.20).

[73] BR112013009157A2 - polipeptídeos com atividade de permease - Google Patents [WWW Document], n.d. URL https://patents.google.com/patent/BR112013009157A2/pt?oq=BR112013009157 (accessed 1.17.20).

[74] WO2019149789A1 - Yeast cell capable of simultaneously fermenting hexose and pentose sugars - Google Patents [WWW Document], n.d. URL https://patents.google.com/patent/WO2019149789A1/en?oq=WO2019%2F149789 (accessed 1.17.20).

[75] WO2017112471A1 - Methods, systems, and compositions for propagation of a fermentation microorganism - Google Patents [WWW Document], n.d. URL https://patents.google.com/patent/WO2017112471A1/en?oq=WO2017%2F112471 (accessed 1.17.20).

[76] WO2010075213A2 - Production of ethanol from lignocellulosic biomass - Google Patents [WWW Document], n.d. URL https://patents.google.com/patent/WO2010075213A2/en?oq=WO2010%2F075213 (accessed 1.17.20).

[77] WO2014045297A2 - A selective microbial production of xylitol from biomass-based sugar stream with enriched pentose component" - Google Patents [WWW Document], n.d. URL https://patents.google.com/patent/WO2014045297A2/en?oq=WO2014%2F045297 (accessed 1.17.20).

[78] WO2010033823A2 - Methods and compositions for degrading pectin - Google Patents [WWW Document], n.d. URL https://patents.google.com/patent/WO2010033823A2/en?oq=WO2010%2F033823 (accessed 1.17.20).

[79] WO2015089428A1 - Enhanced efficiency ethanol production and sugar conversion processes - Google Patents [WWW Document], n.d. URL https://patents.google.com/patent/WO2015089428A1/en?oq=WO201589428 (accessed 1.17.20).

[80] WO2018097844A1 - Methods, systems, and compositions for propagation of a fermentation microorganism - Google Patents [WWW Document], n.d. URL https://patents.google.com/patent/WO2018097844A1/en?oq=WO2018%2F097844 (accessed 1.17.20).

[81] WO2015028583A2 - Glycerol and acetic acid converting cells with improved glycerol transport - Google Patents [WWW Document], n.d. URL https://patents.google.com/patent/WO2015028583A2/en?oq=WO201528583 (accessed 1.17.20).

[82] WO2007005646A9 - Cellules hotes recombinantes et milieu pour production d'ethanol - Google Patents [WWW Document], n.d. URL https://patents.google.com/patent/WO2007005646A9/fr?oq=WO2007%2F005646 (accessed 1.17.20).

[83] WO2010063766A1 - Bioprocessing ligno-cellulose into ethanol with recombinant clostridium - Google Patents [WWW Document], n.d. URL https://patents.google.com/patent/

WO2010063766A1/en?oq=WO2010%2F063766 (accessed 1.17.20).

[84] WO2019058260A1 - Acetate toxicity tolerance in recombinant microbial host cells - Google Patents [WWW Document], n.d. URL https://patents.google.com/patent/WO2019058260A1/en?oq=WO2019%2F058260 (accessed 1.17.20).

[85] Gupta, A, Verma, JP. Sustainable bio-ethanol production from agro-residues: A review. Renewable and Sustainable Energy Reviews. 2025;41:550-567. DOI: 10.1016/j.rser.2014.08.032

[86] Sindhu, R, Binod, P, Pandey, A. Biological pretreatment of lignocellulosic biomass - An overview. Bioresource Technology. 2016;199:76-82. DOI: 10.1016/j.biortech.2015.08.030

[87] Hall, BH, Jaffe, AB, Trajtenberg, M. The NBER Patent-Citations Data File: Lessons, Insights, and Methodological Tools. Patents, citations, and innovations: A window on the knowledge economy. 2001;403. DOI: 10.1186/1471-2164-12-148

[88] Bresnahan, TF, Trajtenberg, M. General purpose technologies "Engines of growth"? Journal of Econometrics. 1995;65(1):83-108. DOI: 10.1016/0304-4076(94)01598-T

[89] OECD. Measuring the Technological and Economic Value of Patents, in: Enquires Into Intellectual Property's Economic Impact. 2015.

[90] Trajtenberg, M, Henderson, R, Jaffe, A. University versus corporate patents: A window on the basicness of invention. Economics of Innovation and New Technology. 1997;5(1):19-50. DOI: 10.1080/10438599700000006

[91] Akinosho, H, Yee, K, Close, D, Ragauskas, A. The emergence of Clostridium thermocellum as a high utility candidate for consolidated bioprocessing applications. Frontiers in Chemistry. 2014;2:66. DOI: 10.3389/fchem.2014.00066

[92] Radecka, D, Mukherjee, V, Mateo, RQ, Stojiljkovic, M, Foulquié-Moreno, MR, Thevelein, JM. Looking beyond *Saccharomyces*: The potential of non-conventional yeast species for desirable traits in bioethanol fermentation. FEMS Yeast Research. 2015;15. DOI: 10.1093/femsyr/fov053

[93] Hasunuma, T, Sanda, T, Yamada, R, Yoshimura, K, Ishii, J, Kondo, A. Metabolic pathway engineering based on metabolomics confers acetic and formic acid tolerance to a recombinant xylose-fermenting strain of Saccharomyces cerevisiae. Microbial Cell Factories. 2011;10(1):2. DOI: 10.1186/1475-2859-10-2

[94] Shen, W, Zhang, J, Geng, B, Qiu, M, Hu, M, Yang, Q, Bao, W, Xiao, Y, Zheng, Y, Peng, W, Zhang, G, Ma, L, Yang, S. Establishment and application of a CRISPR-Cas12a assisted genome-editing system in *Zymomonas mobilis*. Microbial Cell Factories. 2019;18(1):162. DOI: 10.1186/s12934-019-1219-5

[95] Banerjee, S, Mishra, G, Roy, A. Metabolic Engineering of Bacteria for Renewable Bioethanol Production from Cellulosic Biomass. Biotechnology and Bioprocess Engineering. 2019;24(5):713-733. DOI: 10.1007/s12257-019-0134-2

[96] Landhuis, E. Technologies to watch in 2020. Nature. 2020;577:585-587. DOI: 10.1038/d41586-020-00114-4

[97] Buschke, N, Schäfer, R, Becker, J, Wittmann, C. Metabolic engineering of industrial platform microorganisms for biorefinery applications - Optimization of substrate spectrum and process robustness by rational and evolutive strategies. Bioresource Technology. 2013;135:544-554 DOI: 10.1016/j.biortech.2012.11.047

[98] Banerjee, S, Mudliar, S, Sen, R, Giri, B, Satpute, D, Chakrabarti, T, Pandey, RA. Commercializing lignocellulosic bioethanol: Technology bottlenecks and possible remedies. Biofuels, Bioproducts and Biorefining. 2010;4(1):77-93. DOI: 10.1002/bbb.188

[99] Moysés, DN, Reis, VCB, de Almeida, JRM, de Moraes, LMP, Torres, FAG. Xylose fermentation by *saccharomyces cerevisiae*: Challenges and prospects. International Journal of Molecular Sciences. 2016;17(3):207. DOI: 10.3390/ijms17030207

[100] Lee, WH, Jin, YS. Evaluation of ethanol production activity by engineered *Saccharomyces cerevisiae* fermenting cellobiose through the phosphorolytic pathway in simultaneous saccharification and fermentation of cellulose. Journal of Microbiology and Biotechnology. 2017;27(9):1649-1656. DOI: 10.4014/jmb.1705.05039

[101] Bracher, JM, Verhoeven, MD, Wisselink, HW, Crimi, B, Nijland, JG, Driessen, AJM, Klaassen, P, Van Maris, AJA, Daran, JMG, Pronk, JT. The *Penicillium chrysogenum* transporter PcAraT enables high-affinity, glucose-insensitive l-arabinose transport in *Saccharomyces cerevisiae*. Biotechnology for Biofuels. 2018;11(1):63. DOI: 10.1186/s13068-018-1047-6

[102] Ye, S, Jeong, D, Shon, JC, Liu, KH, Kim, KH, Shin, M, Kim, SR. Deletion of PHO13 improves aerobic l-arabinose fermentation in engineered *Saccharomyces cerevisiae*. Journal of Industrial Microbiology and Biotechnology. 2019;46(12):1725-1731. DOI: 10.1007/s10295-019-02233-y

[103] Cunha, JT, Soares, PO, Romaní, A, Thevelein, JM, Domingues, L. Xylose fermentation efficiency of industrial *Saccharomyces cerevisiae* yeast with separate or combined xylose reductase/xylitol dehydrogenase and xylose isomerase pathways. Biotechnology for Biofuels. 2019;12:1-14. DOI: 10.1186/s13068-019-1360-8

[104] Mitsui, R, Yamada, R, Ogino, H. Improved Stress Tolerance of *Saccharomyces cerevisiae* by CRISPR-Cas-Mediated Genome Evolution. Applied Biochemistry and Biotechnology. 2019;10:2637-2650. DOI: 10.1007/s12010-019-03040-y

[105] Sun, J, Wang, J, Tian, K, Dong, Z, Liu, X, Permaul, K, Singh, S, Prior, BA, Wang, Z. A novel strategy for production of ethanol and recovery of xylose from simulated corncob hydrolysate. Biotechnology Letters. 2018a;40:781-788. DOI: 10.1007/s10529-018-2537-0

[106] Sun, J, Tian, K, Wang, J, Dong, Z, Liu, X, Permaul, K, Singh, S, Prior, BA, Wang, Z. Improved ethanol productivity from lignocellulosic hydrolysates by *Escherichia coli* with regulated glucose utilization. Microbial Cell Factories. 2018b;17:1-18. DOI: 10.1186/s12934-018-0915-x

[107] Kim, J, Tremaine, M, Grass, JA, Purdy, HM, Landick, R, Kiley, PJ, Reed, JL. Systems Metabolic Engineering of *Escherichia coli* Improves Coconversion of Lignocellulose-Derived Sugars. Biotechnology Journal. 2019;14. DOI: 10.1002/biot.201800441

[108] Yang, S, Fei, Q, Zhang, Y, Contreras, LM, Utturkar, SM, Brown, SD, Himmel, ME, Zhang, M. *Zymomonas mobilis* as a model system for production of biofuels and biochemicals. Microbial Biotechnology. 2016;9:699-717. DOI: 10.1111/1751-7915.12408

[109] Wang, X, He, Q, Yang, Y, Wang, J, Haning, K, Hu, Y, Wu, B, He, M, Zhang, Y, Bao, J, Contreras, LM, Yang, S. Advances and prospects in metabolic engineering of *Zymomonas mobilis*. Metabolic Engineering. 2018;50:57-73. DOI: 10.1016/j.ymben.2018.04.001

[110] Gonçalves, FA, dos Santos, ES, de Macedo, GR. Alcoholic fermentation of *Saccharomyces cerevisiae*, *Pichia stipitis* and *Zymomonas mobilis* in the presence of inhibitory compounds and seawater. Journal of Basic Microbiology. 2015;55:695-708. DOI: 10.1002/jobm.201400589

[111] Yang, Y, Hu, M, Tang, Y, Geng, B, Qiu, M, He, Q, Chen, S, Wang, X, Yang, S. Progress and perspective on lignocellulosic hydrolysate inhibitor tolerance improvement in *Zymomonas mobilis*. Bioresources and Bioprocessing. 2018;5:1-12. DOI: 10.1186/s40643-018-0193-9

[112] Zhang, K, Lu, X, Li, Y, Jiang, X, Liu, L, Wang, H. New technologies provide more metabolic engineering strategies for bioethanol production in *Zymomonas mobilis*. Applied Microbiology and Biotechnology. 2019;103:2087-2099. DOI: 10.1007/s00253-019-09620-6

[113] Himmel, ME, Ding, SY, Johnson, DK, Adney, WS, Nimlos, MR, Brady, JW, Foust, TD. Biomass recalcitrance: Engineering plants and enzymes for biofuels production. Science. 2007;316:982-982. DOI: 10.1126/science.1137016

[114] Somerville, C, Youngs, H, Taylor, C, Davis, SC, Long, SP. Feedstocks for lignocellulosic biofuels. Science. 2010;329:790-792. DOI: 10.1126/science.1189268

[115] Moreno, AD, Alvira, P, Ibarra, D, Tomás-Pejó, E. Production of Ethanol from Lignocellulosic Biomass. In Production of Platform Chemicals from Sustainable Resources; 2017. p. 375-410. Springer, Singapore. DOI: 10.1007/978-981-10-4172-3_12

[116] Sharma, HK, Xu, C, Qin, W. Biological Pretreatment of Lignocellulosic Biomass for Biofuels and Bioproducts: An Overview. Waste and Biomass Valorization. 2019;10:235-251. DOI: 10.1007/s12649-017-0059-y

[117] Ko, JK, Um, Y, Woo, HM, Kim, KH, Lee, SM. Ethanol production from lignocellulosic hydrolysates using engineered *Saccharomyces cerevisiae* harboring xylose isomerase-based pathway. Bioresource Technology. 2016;209:290-296. DOI: 10.1016/j.biortech.2016.02.124

[118] Wang, L, York, SW, Ingram, LO, Shanmugam, KT. Simultaneous fermentation of biomass-derived sugars to ethanol by a co-culture of an engineered *Escherichia coli* and *Saccharomyces cerevisiae*. Bioresource Technology. 2019;273:269-276. DOI: 10.1016/j.biortech.2018.11.016

[119] Amoah, J, Ogura, K, Schmetz, Q, Kondo, A, Ogino, C. Co-fermentation of xylose and glucose from ionic liquid pretreated sugar cane bagasse for bioethanol production using engineered xylose assimilating yeast. Biomass and Bioenergy. 2019;128:105283. DOI: 10.1016/j.biombioe.2019.105283

[120] Sadalage, PS, Dar, MA, Chavan, AR, Pawar, KD. Formulation of synthetic bacterial consortia and their evaluation by principal component analysis for lignocellulose rich biomass degradation. Renewable Energy. 2019;31:517-535. DOI: 10.1016/j.renene.2019.10.053

[121] Carver, SM, Lepistö, R, Tuovinen, OH. Effect of the type and concentration of cellulose and temperature on metabolite formation by a fermentative thermophilic consortium. International Journal of Hydrogen Energy. 2019;44: 17248-17259. DOI: 10.1016/j.ijhydene.2019.02.177

[122] Zhu, M, Sun, L, Lu, X, Zong, H, Zhuge, B. Establishment of a transient CRISPR-Cas9 genome editing system in *Candida glycerinogenes* for co-production

of ethanol and xylonic acid. Journal of Bioscience and Bioengineering. 2019;128:283-289. DOI: 10.1016/j.jbiosc.2019.03.009

[123] Schultz, JC, Cao, M, Zhao, H. Development of a CRISPR/Cas9 system for high efficiency multiplexed gene deletion in *Rhodosporidium toruloides*. Biotechnology and Bioengineering. 2019;116:2103-2109. DOI: 10.1002/bit.27001

[124] Si, T, Zhao, H. A brief overview of synthetic biology research programs and roadmap studies in the United States. Synthetic and Systems Biotechnology. 2016;1;258-264. DOI: 10.1016/j.synbio.2016.08.003

[125] Cicogna, MPV, Khanna, M, Zilberman, D. Prospects for Biofuel Production in Brazil: Role of Market and Policy Uncertainties. In Handbook of Bioenergy Economics and Policy; 2017. p. 89-117. Springer, New York, NY. DOI: 10.1007/978-1-4939-6906-7_5

[126] Kang, JS, Kholod, T, Downing, S. Analysis of Russia's biofuel knowledge base: A comparison with Germany and China. Energy Policy. 2015;85:182-193. DOI: 10.1016/j.enpol.2015.06.002

[127] MME. RenovaBio [WWW Document]. Petróleo, gás natural e biocombustíveis, 2019.

[128] Portugal-Pereira, J, Soria, R, Rathmann, R, Schaeffer, R, Szklo, A. Agricultural and agro-industrial residues-to-energy: Techno-economic and environmental assessment in Brazil. Biomass and Bioenergy. 2015;81:521-533. DOI: 10.1016/j.biombioe.2015.08.010

[129] Maitan-Alfenas, GP, Visser, EM, Guimarães, VM. Enzymatic hydrolysis of lignocellulosic biomass: Converting food waste in valuable products. Current Opinion in Food Science. 2015;1:44-49. DOI: 10.1016/j.cofs.2014.10.001

[130] Leu, HJ, Wu, CC, Lin, CY. Technology exploration and forecasting of biofuels and biohydrogen energy from patent analysis, in: International Journal of Hydrogen Energy; 2012. p. 15719-15725. DOI: 10.1016/j.ijhydene.2012.04.143

[131] Lin, BW, Chen, WC, Chu, PY. Mergers and acquisitions strategies for industry leaders, challengers, and niche players: Interaction effects of technology positioning and industrial environment. IEEE Transactions on Engineering Management, 2015; 62(1), 80-88. DOI: 10.1109/TEM.2014.2380822

[132] Caviggioli, F, De Marco, A, Scellato, G, Ughetto, E. Corporate strategies for technology acquisition: evidence from patent transactions. Management Decision. 2017;55(6):1163-1181. DOI: 10.1108/MD-04-2016-0220.